Concept Constellations

Propulsion Dynamics and Their Role in Marine Craft Efficiency

Selene Dunlop

Table of Contents

1 Chapter 1: Marine Propulsion Overview

The world of marine propulsion is vast and encompasses an array of technologies, each designed to move vessels through water with varying degrees of efficiency and environmental impact. Understanding the fundamental principles and available options is crucial for anyone involved in the design, operation, or maintenance of marine craft. This chapter serves as a foundational overview, setting the stage for a deeper exploration of specific systems and strategies in the following chapters.

At its core, a marine propulsion system is designed to convert power, typically generated by an engine, into thrust. The journey from power generation to actual movement involves several essential components working in harmony. The engine, whether it's a diesel engine, gas turbine, or electric motor, provides the initial power source. A transmission system, consisting of gears and shafts, transfers this power to the propeller or other propulsor. The propulsor, such as a propeller, waterjet, or even a sail, interacts directly with the water to generate thrust, pushing the vessel forward. Steering mechanisms, like rudders or azimuthing thrusters, control the direction of the vessel. Auxiliary systems, including cooling, lubrication, and fuel delivery systems, are vital for the efficient and reliable operation of the main propulsion components. Each component needs to be designed and integrated carefully to maximize performance and minimize energy losses. A system is only as good as its weakest link, so attention to detail is necessary for all the parts.

Assessing the efficiency of a marine propulsion system is vital for operational cost reduction and environmental stewardship. One key metric is propulsive efficiency, which measures how effectively the power delivered to the propulsor is converted into useful thrust. It considers losses in the transmission system, propeller slip, and other factors. Fuel consumption rate, typically expressed as the amount of fuel consumed per unit of time or distance, is another crucial indicator of efficiency. Specific fuel consumption (SFOC) relates fuel consumption to the power output of the engine, providing a standardized measure for comparison across different engine types and operating conditions. Furthermore, it's critical to consider the overall lifecycle efficiency, accounting for energy inputs during manufacturing, maintenance, and eventual decommissioning. Analyzing these metrics provides a holistic view of a propulsion system's effectiveness and potential for improvement.

Marine propulsion undeniably leaves an environmental footprint, and understanding its impact is paramount. Exhaust emissions from combustion engines, including pollutants like sulfur oxides (SOx), nitrogen oxides (NOx), particulate matter (PM), and carbon dioxide (CO2), contribute to air pollution and climate change.

Regulations such as those from the International Maritime Organization (IMO) are in place to limit these emissions, driving the development of cleaner engine technologies and alternative fuels. Noise pollution from propellers and other machinery can disrupt marine ecosystems, affecting the behavior and communication of marine animals. Oil spills and other forms of pollution associated with maritime activities pose a significant threat to marine life and coastal environments. Sustainable practices, such as using alternative fuels, implementing emission control technologies, and adopting responsible waste management strategies, are essential for minimizing the environmental consequences of marine propulsion. A focus on minimizing the negative impacts of these necessary systems will allow for a greater global conservation effort overall.

Consider, for instance, a container ship relying on a traditional diesel engine powering a fixed-pitch propeller. While this configuration is relatively simple and robust, it can suffer from lower propulsive efficiency at off-design speeds and contribute significantly to air pollution. A modern alternative might involve a dual-fuel engine capable of operating on liquefied natural gas (LNG), which produces fewer emissions than conventional diesel. This engine could be coupled with a controllable-pitch propeller, allowing for more efficient operation across a wider range of speeds. Furthermore, waste heat recovery systems could capture waste heat from the engine exhaust and use it to generate electricity, further improving overall energy efficiency. Finally, using special coatings on the hull will reduce drag and therefore the amount of power to achieve the same goal.

Alternatively, consider a smaller vessel, such as a high-speed ferry. Instead of a propeller, this type of vessel might utilize waterjet propulsion. Waterjets offer several advantages, including high speed and maneuverability, particularly in shallow water. However, they also tend to be less efficient at lower speeds compared to propellers. In this case, the choice of propulsion system depends heavily on the specific operational requirements of the vessel. The speed benefits will make a bigger difference in a small craft like this.

Another interesting avenue is the exploration of electric propulsion systems. While fully electric propulsion is often limited by battery capacity for long-distance voyages, hybrid systems combining electric motors with conventional engines are gaining traction. These hybrid systems can offer improved fuel efficiency, reduced emissions, and quieter operation, especially in port areas. As battery technology continues to advance, we can expect to see a wider adoption of electric and hybrid propulsion in the future. Hybrid systems can have additional benefits, such as power sharing and redundancy.

Ultimately, the selection of a marine propulsion system is a complex decision involving numerous factors, including vessel type, operational profile, environmental regulations, and economic considerations. By understanding the essential components, efficiency metrics, and environmental impacts associated with different propulsion methods, we can make informed choices that optimize performance, minimize costs, and protect our marine ecosystems. As technology evolves and environmental awareness grows, the pursuit of more efficient and sustainable marine propulsion solutions will continue to drive innovation in the maritime industry.

1.1 Propulsion System Essentials

At its core, a marine propulsion system serves the fundamental purpose of generating thrust to overcome resistance and propel a vessel through the water. Understanding the essential components and their interplay is crucial for anyone involved in naval architecture, marine engineering, or vessel operation. The conventional system generally comprises an engine, a transmission system, and a propeller, each playing a vital, distinct role.

The engine, often a diesel engine due to its high torque and fuel efficiency, acts as the prime mover, converting chemical energy into mechanical energy. However, steam turbines and gas turbines are also common depending on the application, such as larger vessels, where power demands are higher. Diesel engines, in particular, are favored for their robustness, reliability, and ability to burn heavy fuel oil, a relatively inexpensive option. The power output of the engine must be carefully matched to the vessel's size, displacement, and intended operating speed. Insufficient power will lead to poor performance, while excessive power results in unnecessary fuel consumption and increased wear and tear. Choosing the right engine also involves considering factors such as emissions regulations, maintenance requirements, and lifecycle costs. Modern engines often incorporate advanced technologies such as electronic fuel injection, turbocharging, and aftercooling to optimize performance and minimize emissions. Furthermore, advancements in engine design have led to improved fuel efficiency and reduced noise levels.

Following the engine is the transmission system, which serves to transfer the mechanical power from the engine to the propeller. This often involves a gearbox, which adjusts the engine's high rotational speed to a more suitable speed for the propeller. Direct drive systems exist but are less common in modern vessels due to the limitations they impose on engine speed and propeller efficiency. The gearbox ratio is a crucial parameter, selected to optimize the propeller's performance at the vessel's desired cruising speed. In some cases, the transmission system also incorporates a clutch, which allows the engine to be disconnected from the propeller, enabling maneuvering and station keeping. Hydraulic couplings or electric drives can also be implemented as part of the transmission system, providing smoother power transmission and improved control. For larger vessels with multiple engines, complex transmission systems may be used to combine the power output of several engines to drive a single propeller shaft. Such systems require careful design and synchronization to ensure efficient and reliable operation.

The propeller, submerged and rotating, converts the rotational mechanical energy into thrust, accelerating a mass of water to propel the vessel forward. The propeller's design characteristics, such as its diameter, pitch, and blade area ratio, significantly influence its performance. A larger diameter propeller generally provides more thrust at lower speeds, while a smaller diameter propeller is more efficient at higher speeds. The pitch, which is the distance a propeller would advance in one revolution if there were no slip, determines the propeller's thrust-generating capability. The blade area ratio affects the propeller's ability to absorb power and resist cavitation, a phenomenon where vapor bubbles form on the propeller blades, reducing efficiency and causing damage. Propeller design is a complex process that involves balancing these factors to optimize performance for a specific vessel and operating conditions. Different propeller types exist, each with its own advantages and disadvantages. Fixed-pitch propellers are simple and robust but offer limited flexibility in terms of adjusting thrust and speed. Controllable-pitch propellers allow the pitch angle to be adjusted while the engine is running, providing greater control over thrust and speed. Ducted propellers, enclosed within a nozzle, can improve thrust and efficiency, particularly at lower speeds.

Beyond the core components, auxiliary systems play a crucial role in supporting the propulsion system's operation. These include cooling systems, lubricating systems, fuel systems, and exhaust systems. The cooling system removes heat generated by the engine and other components, preventing overheating and ensuring reliable operation. The lubricating system reduces friction between moving parts, minimizing wear and tear and extending the lifespan of the equipment. The fuel system supplies fuel to the engine, ensuring a consistent and reliable fuel supply. The exhaust system removes combustion gases from the engine, minimizing emissions and noise pollution. These auxiliary systems are essential for maintaining the propulsion system's overall performance and reliability. Proper maintenance and monitoring of these systems are crucial for preventing breakdowns and ensuring safe operation.

Navigating the complexities of propulsion systems requires a keen understanding of various factors beyond just the mechanics. Materials play a vital role, especially in seawater environments. Selecting corrosion-resistant alloys for propellers, shafts, and other submerged components is essential to ensure longevity and prevent premature failure. Stainless steels, bronzes, and nickel-aluminum bronzes are commonly used materials, each with its own specific properties and applications. The choice of material depends on factors such as the vessel's operating environment, the propeller's design, and the expected service life. Furthermore, the manufacturing process used to produce these components can also affect their performance and durability. Casting, forging, and machining are common manufacturing methods, each with its own advantages and limitations.

Maintenance is paramount for any propulsion system. Regular inspections, lubrication, and component replacements are essential for maintaining optimal performance and preventing costly breakdowns. A well-defined maintenance schedule, tailored to the specific propulsion system and operating conditions, is crucial. This schedule should include regular checks of fluid levels, filter replacements, and visual inspections of critical components. Furthermore, vibration analysis and oil analysis can be used to detect potential problems early on, allowing for proactive maintenance and preventing catastrophic failures. Proper maintenance not only extends the lifespan of the propulsion system but also improves its efficiency and reduces its environmental impact.

1.2 Efficiency Metrics in Propulsion

Quantifying the effectiveness of a marine propulsion system goes far beyond simply measuring speed. It involves a deep dive into various efficiency metrics that offer a comprehensive understanding of how well a system converts fuel energy into useful thrust while minimizing losses. These metrics provide crucial insights for design optimization, performance monitoring, and ultimately, reducing operational costs and environmental impact. Without these, we would be making decisions blindfolded.

One of the most fundamental metrics is propulsive efficiency (p). This value represents the ratio of useful power delivered to the vessel (thrust power) to the power actually delivered by the propulsion system (delivered power). Mathematically, p = PT / PD, where PT is the thrust power (the product of thrust and vessel speed) and PD is the power delivered to the propeller or propulsor shaft. Achieving a high propulsive efficiency requires careful matching of the propeller to the hull and engine characteristics, as well as minimizing losses due to wake, thrust deduction, and relative rotative efficiency. For example, a vessel designed for slow, steady cruising will benefit from a large diameter, slow-turning propeller optimized for high thrust at low speeds, while a high-speed vessel will require a smaller diameter, faster-turning propeller designed to minimize cavitation and maximize thrust at higher speeds. Furthermore, the hull shape must be carefully designed to minimize resistance and provide a uniform inflow of water to the propeller.

Next up is quasi-propulsive coefficient (QPC), offering a more holistic view of the propulsion system's efficiency, including hull interaction effects. Unlike propulsive efficiency which focuses solely on the propeller's performance, QPC considers the entire propulsion chain from engine output to effective tow power. The QPC is the ratio of effective tow power (PE) to the delivered power (PD). The effective tow power is the power required to tow the hull at a given speed in open water. This means, QPC = PE/PD. Since PE = RT * VS where RT is the total resistance and VS is the ship speed. A higher QPC indicates a more efficient system, taking into account the interaction between the hull and the propulsor. Factors that influence QPC include hull form, propeller design, and the presence of appendages. Consider the placement of rudders and stabilizers; their alignment and design can significantly impact the flow of water around the hull, either enhancing or diminishing the effectiveness of the propeller. Careful hydrodynamic analysis is crucial to optimize these elements for maximum QPC.

Speaking of fuel consumption, specific fuel consumption (SFC) is a critical metric, particularly in today's world where both cost and environmental regulations are paramount. SFC quantifies the amount of fuel consumed per unit of power produced, typically measured in grams per kilowatt-hour (g/kWh) or pounds per horsepower-hour (lb/hp-hr). A lower SFC indicates greater fuel efficiency. Modern engine designs focus on minimizing SFC through improved combustion processes, reduced friction, and optimized operating parameters. The choice of fuel itself also plays a significant role; for instance, liquefied natural gas (LNG) and other alternative fuels often offer lower SFC compared to traditional heavy fuel oil (HFO). Beyond engine technology, operational practices also have a substantial impact on SFC. Regular maintenance, proper trim and ballast management, and optimized route planning can all contribute to significant fuel savings. Weather routing, for example, allows vessels to avoid adverse sea conditions, reducing resistance and fuel consumption.

Moving to a more specific element, propeller open water efficiency (o) is an important metric as well. This measures the efficiency of the propeller operating in isolation, without the influence of the hull. It is defined as the ratio of thrust power to input power (torque times rotational speed) of the propeller in open water conditions. This metric is primarily used during the propeller design phase to assess the intrinsic performance of the propeller itself. It helps engineers optimize propeller geometry, blade pitch, and other parameters to maximize its thrust-producing capability. However, it's crucial to remember that o does not fully represent the propeller's performance when installed on a vessel, as it doesn't account for hull-propeller interaction effects. Nevertheless, a high o is a good starting point for achieving overall high propulsive efficiency.

Beyond those already mentioned, we can look at the cavitation number (). This dimensionless number is an important indicator of the propensity of a propeller to experience cavitation. Cavitation occurs when the local pressure around the propeller blades drops below the vapor pressure of water, leading to the formation of vapor bubbles. These bubbles collapse violently, causing noise, erosion, and reduced thrust. The cavitation number is defined as = (P \item Pv) / (0.5 \emph{ } V^2), where P is the static pressure, Pv is the vapor pressure of water, is the density of water, and V is a characteristic velocity (usually the propeller tip speed). A lower cavitation number indicates a higher risk of cavitation. Propeller designs aim to maintain a high cavitation number by optimizing blade shapes and operating conditions. Furthermore, the depth of the propeller below the water surface is a critical factor in preventing cavitation, as increased depth leads to higher static pressure.

Finally, understanding wake fraction (w) and thrust deduction fraction (t) is vital for understanding hull-propeller interaction. The wake fraction (w) represents the reduction in water velocity experienced by the propeller due to the presence of the hull. The propeller operates in the "wake" of the hull, which has a lower velocity than the freestream velocity. The thrust deduction fraction (t) accounts for the increase in resistance experienced by the hull due to the operation of the propeller. The propeller suction pulls on the hull, increasing its resistance. These fractions are used to correct the open water propeller characteristics to account for the actual operating conditions behind the hull. Accurately determining these fractions is crucial for predicting the overall propulsive performance of the vessel. Model testing and computational fluid dynamics (CFD) simulations are commonly used to estimate wake fraction and thrust deduction fraction.

1.3 Environmental Impact Considerations

Environmental Impact Considerations
The impact of marine propulsion systems extends far beyond the immediate operational costs and efficiency ratings. Every decision, from the choice of fuel to the design of the propeller, carries a ripple effect through the marine environment. For beginners in the field, grasping these interconnected consequences is not merely academic; it is fundamental to designing and operating vessels responsibly. We will explore the tangible implications of propulsion choices on our oceans, focusing on pollutants, noise, and physical disturbances.
The issue of emissions looms large when discussing environmental impact. Traditional marine fuels, primarily heavy fuel oil (HFO) and marine diesel oil (MDO), release significant quantities of air pollutants. Sulfur oxides (SOx), nitrogen oxides (NOx), particulate matter (PM), and carbon dioxide (CO2) are among the most concerning. SOx emissions contribute to acid rain and respiratory problems, while NOx contributes to smog and acid deposition. PM, especially the ultra-fine particles, poses severe health risks to both humans and marine life. CO2, of course, is a primary driver of climate change and ocean acidification. Regulations such as those established by the International Maritime Organization (IMO) through MARPOL Annex VI are pushing for lower sulfur content in fuels and setting limits on NOx emissions. Understanding these regulations and the technologies available to meet them, such as scrubbers for SOx removal and selective catalytic

reduction (SCR) for NOx reduction, is a crucial first step towards environmentally conscious propulsion. It's not just about adhering to the rules; it's about actively seeking cleaner alternatives. The initial cost of these technologies can seem daunting, but failing to address emissions now will inevitably lead to higher costs later, both financially and environmentally.

Beyond air pollution, the underwater noise generated by ships presents a less visible, yet equally damaging, environmental problem. Propellers, particularly those experiencing cavitation, are a major source of this noise. Marine mammals, such as whales and dolphins, rely heavily on acoustic communication for navigation, foraging, and social interaction. Excessive noise can disrupt these activities, leading to stress, disorientation, and even physical harm. Studies have shown that increased shipping noise is linked to changes in whale migration patterns, reduced foraging success, and increased strandings. Fish populations are also affected by underwater noise, with studies showing impacts on spawning behavior and larval development. Quieter propeller designs, such as skewed propellers and those operating at lower speeds, can significantly reduce noise emissions. Hull design also plays a role; streamlining the hull and reducing turbulence can minimize noise generated by water flow. Consider this: a seemingly minor adjustment to propeller geometry can have a profound impact on the well-being of marine animals hundreds of miles away. As future engineers and designers, we need to prioritize noise reduction as a key design parameter.

The physical impact of marine propulsion extends to the direct disturbance of marine habitats. Propellers can cause erosion and sedimentation, particularly in shallow waters and near coastlines. The wake generated by high-speed vessels can damage sensitive ecosystems, such as coral reefs and seagrass beds. Invasive species, transported in ballast water or attached to the hull, can wreak havoc on local biodiversity. Thoughtful route planning and speed management can minimize these physical disturbances. Avoiding sensitive areas, reducing speed in shallow waters, and implementing ballast water treatment systems are all essential steps. I remember once working on a project where the proposed route of a new ferry line would have directly impacted a known seagrass bed. By rerouting the ferry and reducing its speed in the area, we were able to minimize the environmental impact and protect the delicate ecosystem. It highlighted the importance of considering these physical impacts early in the design process.

Moving forward, it's important to emphasize the adoption of sustainable fuel alternatives. Liquefied Natural Gas (LNG) is increasingly being used as a transition fuel, offering lower SOx, NOx, and PM emissions compared to traditional fuels. However, LNG is still a fossil fuel and contributes to greenhouse gas emissions. Biofuels, produced from renewable sources such as algae or waste biomass, offer a more sustainable option, provided they are produced in an environmentally responsible manner. Hydrogen fuel cells, powered by hydrogen produced from renewable energy sources, hold immense promise for zero-emission propulsion. Battery-electric propulsion is also gaining traction, particularly for short-range vessels and hybrid systems. Remember, the transition to sustainable fuels will not happen overnight. It requires investment in infrastructure, technological development, and supportive policies. Yet, the potential benefits - cleaner air, healthier oceans, and a more sustainable future - are well worth the effort.

Consider life cycle assessment (LCA) as a tool for evaluating the environmental footprint of marine propulsion systems. LCA considers the entire life cycle of a product or system, from raw material extraction to end-of-life disposal. This includes energy consumption, emissions, and resource use at each stage. By conducting an LCA, you can identify the most environmentally impactful aspects of a propulsion system and focus on areas for improvement. For instance, an LCA might reveal that the manufacturing process of a particular type of propeller has a significantly higher environmental footprint than the operation of the propeller itself. This would prompt you to investigate alternative manufacturing processes or materials. It provides a comprehensive framework for understanding the true environmental cost of our choices.

Embracing a holistic approach to environmental impact considerations is a responsibility we all share. This goes beyond simply complying with regulations. It requires continuous learning, innovation, and a commitment to minimizing our footprint on the marine environment. The choices we make today will shape the health of our oceans for generations to come. As a new entrant to the marine propulsion field, you are not just designing and building ships; you are shaping the future of our planet.

2 Chapter 2: Hydrodynamic Principles

Understanding the forces at play when a vessel moves through water is paramount to designing efficient and effective marine craft. We must first consider the very nature of the fluid itself. Water, unlike air, is relatively incompressible, meaning its density remains largely constant under pressure changes encountered in most marine applications. This characteristic influences how water flows around a hull and how it interacts with propulsion devices. Viscosity, a measure of a fluid's resistance to flow, is another critical property. A high viscosity fluid, like honey, resists movement more than a low viscosity fluid, like water. Seawater's viscosity affects the frictional resistance experienced by a hull, which is directly linked to fuel consumption. Beyond these properties, the concept of pressure is also critical; vessels displace water, creating pressure differentials. These pressure variations are key to understanding lift generation on hydrofoils or the suction forces around a propeller.

Considering the flow of water around a submerged or semi-submerged body, there are two primary flow regimes: laminar and turbulent. Laminar flow is characterized by smooth, orderly layers of fluid moving parallel to each other. It is often found very close to the surface of a vessel, in a region called the boundary layer. Turbulent flow, on the other hand, is chaotic and irregular, with swirling eddies and fluctuations in velocity. Most of the flow around a marine vessel, especially at higher speeds, is turbulent. These different flow regimes have distinct effects on resistance. Laminar flow generates less friction than turbulent flow, but turbulent flow is generally more resistant to separation, where the flow detaches from the surface of the body. Understanding these flow regimes allows naval architects to design hull forms that minimize turbulence and delay flow separation, resulting in reduced drag and increased efficiency. Moreover, you must consider the Reynolds number, a dimensionless quantity that helps predict whether flow will be laminar or turbulent. A high Reynolds number typically indicates turbulent flow.

Moving from these basics, we encounter the forces that impede a vessel's motion, primarily resistance and drag. Resistance, in a marine context, is the sum of all forces opposing the movement of a vessel through water. This overall resistance can be divided into several key components. Frictional resistance, already mentioned, arises from the friction between the hull surface and the water. It is directly proportional to the wetted surface area of the hull, the square of the vessel's speed, and the viscosity of the water. Wave-making resistance is another significant component, particularly at higher speeds. As a vessel moves, it creates waves on the water surface, expending energy in the process. The magnitude of wave-making resistance depends on the vessel's speed, length, and hull form. A longer vessel generally creates smaller waves, and therefore experiences less wave-making resistance at a given speed. Form drag, or pressure drag, arises from pressure differences created as water flows around the hull. A streamlined hull form minimizes form drag by reducing pressure variations. Air resistance, while often a smaller component, becomes relevant at higher speeds and for vessels with significant above-water structures. Appendage resistance is the drag created by appendages such as rudders, stabilizers, and sonar domes. Finally, added resistance in waves is caused by the vessel encountering waves, either in head seas or following seas.

Quantifying these resistance components is critical for predicting vessel performance and optimizing hull designs. Several methods are used to estimate resistance. Empirical formulas, based on experimental data from model testing, provide estimates for specific hull forms and operating conditions. These formulas often include coefficients that account for the effects of hull shape, speed, and water conditions. Computational Fluid Dynamics (CFD) simulations offer a more detailed and accurate approach. CFD involves solving the governing equations of fluid flow using numerical methods. CFD simulations can predict the flow field around a vessel, allowing engineers to visualize and quantify resistance components. Model testing, in towing tanks or circulating water channels, remains a valuable tool for validating CFD results and obtaining data for empirical formulas. Scale models of vessels are tested under controlled conditions, and the measured resistance is scaled up to predict the full-scale performance.

Optimizing hull form to minimize resistance is a crucial aspect of naval architecture. Streamlining the hull shape is essential to reduce form drag and promote laminar flow. Bulbous bows, strategically positioned below the waterline, can reduce wave-making resistance by creating a wave system that interferes destructively with the bow wave. Stern wedges and flaps can improve flow around the stern, reducing drag and increasing propulsion efficiency. Hull coatings can reduce frictional resistance by creating a smoother surface. These coatings can range from traditional paints to advanced fouling-release coatings that prevent marine organisms from attaching to the hull. Air lubrication systems inject air bubbles beneath the hull, reducing frictional resistance by creating a layer of air between the hull and the water. Every choice, from the overall hull shape to the surface finish, plays a role in determining the overall resistance and efficiency of the vessel.

In conclusion, hydrodynamic principles are fundamental to understanding and optimizing the performance of marine craft. A thorough understanding of fluid dynamics, resistance components, and optimization techniques is essential for designing efficient and effective vessels. By applying these principles, naval architects and engineers can reduce fuel consumption, increase speed, and minimize the environmental impact of marine transportation.

2.1 Fluid Dynamics Basics

Fluid dynamics, at its core, is the study of how fluids - liquids and gases - behave when in motion. It's a cornerstone of understanding marine propulsion, as water is the medium through which our vessels move and interact. Without a solid grasp of these basic principles, optimizing a boat's design for efficiency becomes an exercise in guesswork. Let's dive into these fundamentals.

One of the first concepts you'll encounter is viscosity. Think of it as a fluid's resistance to flow; honey has high viscosity, while water has relatively low viscosity. This property is fundamentally important because it dictates how much energy is required to move a fluid around an object, or conversely, to move an object through a fluid. High viscosity translates to greater frictional resistance. When a hull moves through water, the water molecules closest to the hull surface experience more friction than those further away, creating what we call a boundary layer.

Pressure is another crucial element. It's defined as force per unit area, and in fluid dynamics, we often deal with pressure gradients - differences in pressure between two points. These gradients drive fluid flow. A high-pressure area will naturally push fluid towards a low-pressure area. This is the underlying principle behind many propulsion systems. The shape of a hull and propeller blades are meticulously designed to create specific pressure distributions around them, driving the vessel forward. Understanding how pressure varies around a hull allows naval architects to minimize areas of high pressure (which lead to drag) and maximize areas of low pressure in strategic locations (which contribute to thrust).

Now, consider the concept of flow. Fluid flow can be broadly classified into two categories: laminar and turbulent. Laminar flow is characterized by smooth, orderly movement of fluid particles in layers, with minimal mixing. Turbulent flow, on the other hand, is chaotic and irregular, with swirling eddies and significant mixing. While laminar flow is generally desirable because it minimizes energy loss due to friction, it's often difficult to achieve in real-world applications, especially around complex hull shapes at high speeds. The transition from laminar to turbulent flow is influenced by factors like fluid velocity, viscosity, and the geometry of the object the fluid is flowing around. To better understand the transition it's beneficial to consider the Reynolds number. This dimensionless number helps predict whether flow conditions will be laminar or turbulent and is calculated as the ratio of inertial forces to viscous forces. Higher Reynolds numbers generally indicate a greater tendency towards turbulence.

Bernoulli's principle offers another lens through which to view fluid dynamics. It states that for an inviscid flow (one with negligible viscosity), an increase in the speed of the fluid occurs simultaneously with a decrease in pressure or a decrease in the fluid's potential energy. In simpler terms, faster-moving fluid has lower pressure. This principle is readily observed in the design of airfoils and, relevantly, propeller blades. The curved shape of a blade causes the water to travel faster over the upper surface than the lower surface, resulting in a pressure difference that generates lift (or, in the case of a propeller, thrust).

Drag is a resistive force that opposes the motion of an object through a fluid. As mentioned before, it's split into several components. Frictional drag is directly related to the viscosity of the fluid and the surface area of the object in contact with the fluid. Wave-making drag, occurs when a vessel moves through water, generating waves that radiate away from the hull, consuming energy in the process. Form drag, also known as pressure drag, arises from the pressure difference between the front and rear of an object. Streamlining the hull can significantly reduce form drag by allowing the water to flow smoothly around the vessel, minimizing pressure imbalances.

Finally, a crucial understanding is the concept of boundary layers, those thin layers of fluid near a surface that are significantly affected by friction. Understanding the nature of the boundary layer is essential for predicting and controlling drag. These layers are very thin, but have a huge impact on the overall resistance a hull faces. Engineers are constantly seeking ways to manipulate the boundary layer, encouraging it to remain laminar for as long as possible, to reduce friction.

In conclusion, mastery of these fundamental fluid dynamics principles is not merely theoretical; it's intensely practical. By understanding viscosity, pressure, flow regimes, Bernoulli's principle, drag components, and boundary layers, we can begin to appreciate the intricate dance between a marine craft and the water it moves through. This understanding is the bedrock upon which all subsequent design and efficiency improvements are built.

2.2 Resistance and Drag Forces

Understanding the forces that impede a vessel's movement through water is crucial for designing efficient and effective marine craft. Resistance and drag, often used interchangeably, encompass a complex interplay of hydrodynamic phenomena that dictate the power required to achieve a desired speed. For anyone looking to minimize fuel consumption and maximize performance, grasping these concepts is not just academic, it's fundamental. Let's break down these forces, their origins, and how they impact your designs or operational decisions.

Frictional resistance is the first hurdle to overcome. It arises from the viscous friction between the hull surface and the water flowing past it. Think of it as the water molecules clinging to the hull, creating a drag force that opposes motion. The magnitude of this force is directly proportional to the wetted surface area of the hull, the square of the vessel's speed, and the viscosity of the water. A larger, rougher hull experiences significantly more frictional resistance than a smaller, smoother one. Factors such as biofouling - the accumulation of marine organisms on the hull - drastically increase frictional resistance, leading to substantial fuel penalties. Therefore, regular hull cleaning and the application of specialized antifouling coatings are vital practices for maintaining optimal performance. I have personally witnessed the dramatic impact of even a thin layer of slime on a vessel's fuel consumption during sea trials, highlighting the importance of proactive maintenance. Different types of coatings offer varying degrees of protection and longevity, with newer technologies focusing on non-toxic alternatives to minimize environmental impact. The specific choice of coating should be tailored to the vessel's operating profile, considering factors like speed, water temperature, and typical docking locations.

Wave-making resistance is another major component of total resistance, particularly at higher speeds. As a vessel moves through water, it creates waves. The energy required to generate these waves is drawn from the vessel's propulsive power, effectively acting as a drag force. Wave-making resistance is highly dependent on the Froude number, a dimensionless parameter that relates the vessel's speed to its length. At low Froude numbers, wave-making resistance is relatively small, but as speed increases, it rises dramatically, often becoming the dominant component of total resistance. This is because the wavelength of the waves generated by the vessel becomes comparable to the vessel's length, leading to significant energy dissipation. Hull form plays a crucial role in mitigating wave-making resistance. Slender, streamlined hulls with fine bows tend to generate smaller waves than bluff, boxy hulls. The bulbous bow, a characteristic feature of many modern ships, is specifically designed to reduce wave-making resistance by creating a wave system that interferes destructively with the bow wave. Through careful hydrodynamic design, engineers can minimize the energy wasted in wave generation, leading to significant improvements in fuel efficiency.

Form resistance, also known as pressure resistance, is a component of resistance that arises from the pressure distribution around the hull. As water flows around the hull, it accelerates and decelerates, creating regions of high and low pressure. If the hull is not streamlined, these pressure differences can lead to significant drag. Bluff bodies, for example, experience high form resistance due to the large pressure difference between the front and rear surfaces. Streamlining the hull minimizes these pressure differences, reducing form resistance. This is why submarines and racing yachts are designed with highly streamlined shapes. Appendages such as rudders, keels, and struts also contribute to form resistance, so their design and placement must be carefully considered. Fairing these appendages, by adding streamlined shapes to their leading and trailing edges, can significantly reduce their contribution to form resistance. Even small details, such as the shape of the rudder stock or the design of the propeller hub, can have a noticeable impact on overall resistance. Air resistance, while often overlooked, can become significant at higher speeds, especially for vessels with large superstructures. The force of the wind against the vessel's above-water surfaces creates a drag that opposes motion. The magnitude of air resistance depends on the vessel's frontal area, the wind speed, and the shape of the superstructure. Streamlining the superstructure, by rounding corners and minimizing sharp edges, can reduce air resistance. However, the primary strategy for minimizing air resistance is often to reduce the exposed area as much as possible, while still maintaining functionality and safety. This may involve optimizing the placement of antennas, masts, and other equipment, and designing the superstructure to be as compact and aerodynamic as possible. Understanding the relative importance of air resistance is particularly important for high-speed craft and vessels operating in windy conditions.

Finally, added resistance due to waves is a factor that must be accounted for in real-world operating conditions. When a vessel encounters waves, it experiences additional resistance beyond that encountered in calm water. This added resistance arises from several factors, including wave reflection, diffraction, and the vessel's motions in response to the waves. The magnitude of added resistance depends on the wave height, wavelength, and direction, as well as the vessel's size, shape, and speed. In head seas (waves approaching from the front), added resistance can be particularly significant, as the vessel repeatedly encounters and deflects oncoming waves. In following seas (waves approaching from behind), the added resistance may be less pronounced, but the vessel can experience other challenges, such as broaching and surfing. Predicting and mitigating added resistance due to waves is a complex task that requires sophisticated hydrodynamic analysis. However, even a basic understanding of the underlying principles can help operators make informed decisions about speed and course adjustments to minimize fuel consumption and maintain safe operation.

Ultimately, minimizing resistance and drag forces is a multifaceted challenge that requires a holistic approach. It involves careful consideration of hull form, surface roughness, appendage design, superstructure streamlining, and operational practices. By understanding the underlying principles and employing appropriate strategies, it is possible to significantly improve the efficiency and performance of marine craft, reducing fuel consumption, lowering emissions, and enhancing overall operational effectiveness. The interplay of these factors is what makes marine design so challenging and rewarding.

3 Chapter 3: Propeller Design and Analysis

Propeller design is as much art as it is science, a delicate balance between theoretical calculations and practical considerations. The goal, simply put, is to convert the rotational power of the engine into thrust, moving the vessel efficiently through the water. This conversion, however, is rife with challenges, ranging from cavitation to hydrodynamic losses. Understanding these challenges and mastering the tools to mitigate them are essential for any aspiring naval architect or marine engineer.

Let's dive into propeller geometry. The diameter, the overall size of the propeller disk, dictates the amount of water the propeller can act upon. A larger diameter generally results in higher thrust at lower speeds, crucial for applications like tugboats or vessels operating in confined waters. The pitch, on the other hand, represents the distance a propeller would advance in one revolution if it were moving through a solid medium. It's a measure of the helix angle of the blades, directly influencing the speed at which the vessel can travel. A higher pitch translates to a higher potential speed, but it requires more power to overcome the increased resistance. Blade area ratio (BAR) is another critical parameter, defined as the ratio of the projected blade area to the area of the propeller disk. A higher BAR provides greater thrust capacity and reduces the risk of cavitation, but it also increases drag. Furthermore, the number of blades impacts both performance and vibration characteristics. More blades generally result in smoother operation but can also increase drag and complexity. The interplay between these geometric parameters is complex, demanding careful consideration to optimize performance for specific operational profiles.

Blade Element Theory (BET) is the bedrock of propeller analysis. It's a method that divides the propeller blade into a series of independent elements, each treated as a two-dimensional airfoil. This allows engineers to calculate the lift and drag forces acting on each element, and then integrate these forces along the entire blade to determine the overall thrust and torque produced by the propeller. The beauty of BET lies in its simplicity and adaptability. It accounts for factors such as the angle of attack, the relative velocity of the water, and the airfoil characteristics of the blade section. However, it's important to recognize the limitations of BET. It neglects three-dimensional flow effects, such as tip vortices and radial flow, which can significantly influence propeller performance, particularly at high speeds or heavily loaded conditions. Despite these limitations, BET provides a valuable starting point for propeller design and analysis, serving as a foundation for more sophisticated computational fluid dynamics (CFD) simulations.

Cavitation, the formation and implosion of vapor bubbles on the propeller blades, is the bane of propeller designers. It occurs when the local pressure on the blade surface drops below the vapor pressure of the water, causing the water to boil and form bubbles. These bubbles collapse violently as they move into regions of higher pressure, generating noise, vibration, and erosion damage to the propeller blades. Cavitation not only reduces propeller efficiency but can also significantly shorten the lifespan of the propeller. To mitigate cavitation, designers employ a variety of strategies. Increasing the blade area ratio can reduce the load on each blade element, lowering the pressure drop and delaying the onset of cavitation. Optimizing the blade section shape to minimize pressure peaks is also crucial. Furthermore, controlling the propeller's operating depth and avoiding high rotational speeds can help to prevent cavitation. Predicting cavitation accurately requires sophisticated CFD simulations, capable of capturing the complex interplay between pressure, velocity, and vapor formation.

Predicting propeller performance is a crucial step in the design process, allowing engineers to assess the suitability of a propeller for a specific vessel and operating conditions. Open-water tests, conducted in towing tanks or cavitation tunnels, provide valuable experimental data on propeller performance under controlled conditions. These tests measure thrust, torque, and rotational speed, allowing for the determination of propeller efficiency and other performance parameters. Empirical formulas, derived from experimental data, offer a quick and convenient way to estimate propeller performance, particularly during the initial design stages. These formulas typically relate propeller parameters, such as diameter, pitch, and blade area ratio, to performance characteristics, such as thrust coefficient and torque coefficient. Computational Fluid Dynamics (CFD) simulations provide the most detailed and accurate method for predicting propeller performance. CFD simulations solve the Navier-Stokes equations, which govern fluid flow, to simulate the complex hydrodynamic phenomena around the propeller. These simulations can capture three-dimensional flow effects, cavitation, and other complex phenomena that are not accounted for by simpler methods.

Beyond conventional propellers, advanced propeller concepts are constantly being developed to improve efficiency and reduce noise. Ducted propellers, also known as Kort nozzles, surround the propeller with a duct, which accelerates the flow of water and increases thrust. Ducted propellers are particularly effective at low speeds, making them suitable for tugboats and other vessels requiring high bollard pull. Contrarotating propellers consist of two propellers rotating in opposite directions on the same axis. This configuration recovers some of the energy lost in the propeller slipstream, increasing efficiency and reducing noise. Overlapping Propellers is another advanced concept, using two propellers in partial overlap and increasing the efficiency. Azimuthing propellers, or Z-drives, combine propulsion and steering in a single unit, allowing for 360-degree thrust vectoring. This provides excellent maneuverability and is particularly useful for dynamic positioning and harbor maneuvering. These advanced concepts represent the cutting edge of propeller technology, pushing the boundaries of efficiency and performance.

As a beginner, diving into the intricacies of propeller design can feel overwhelming. Start with the fundamentals - understand the geometry, grasp the principles of Blade Element Theory, and learn about cavitation. Explore the different methods for performance prediction, from empirical formulas to CFD simulations. Don't be afraid to experiment and iterate, using software tools and online resources to explore different design options. The field of marine propulsion is constantly evolving, with new technologies and concepts emerging all the time. Stay curious, keep learning, and never stop exploring the fascinating world of propeller design. Remember, the most effective designs are the ones that strike a balance between theoretical elegance and practical considerations, resulting in efficient and reliable propulsion systems that power vessels across the globe.

3.1 Propeller Geometry and Terminology

Let's embark on a journey through the intricate world of propeller geometry, a cornerstone for understanding and optimizing marine propulsion. The propeller, seemingly simple in its function, relies on a complex interplay of meticulously designed geometric parameters. Each curve, each angle, each dimension contributes to its overall efficiency and performance. Grasping this terminology is paramount for anyone involved in the design, operation, or maintenance of marine vessels.

The diameter, represented by 'D', immediately jumps out as the most fundamental characteristic. It is simply the distance from one blade tip to the opposite blade tip, measured across the circle swept by the propeller. Think of it as the wingspan of our underwater bird. A larger diameter usually translates to a greater thrust potential, but also higher torque requirements. Conversely, a smaller diameter offers maneuverability benefits in shallow waters, but it may compromise thrust generation. Therefore, the diameter is carefully selected based on the vessel size, hull form, engine power, and intended operating conditions. It's a balancing act, a compromise between power and practicality.

Following that, we must consider pitch, which is the theoretical distance a propeller advances forward in one complete revolution if it were moving through a solid medium. It's typically denoted by 'P'. Essentially, it describes the angle of the propeller blades; a higher pitch means a steeper angle. A high pitch propeller is suited for vessels requiring high speed, while a lower pitch is more suitable for applications requiring greater thrust at lower speeds, such as towing or maneuvering heavy loads. However, the actual distance a propeller advances (the 'advance') is always less than the pitch due to slip.

Slip is inevitable and represents the difference between the theoretical advance (pitch) and the actual advance per revolution. It arises from the fact that water is not a solid medium, and the propeller blades create a wake, reducing the effective forward motion. Slip is expressed as a percentage and is an important indicator of propeller efficiency. Higher slip values typically indicate lower efficiency and can result from factors such as excessive loading or incorrect propeller selection.

The blade number, denoted as 'Z', is quite simply the number of blades on the propeller. While a two-bladed propeller might be suitable for small boats, larger vessels commonly employ three, four, or even five-bladed propellers. Increasing the number of blades generally enhances thrust and reduces vibration, but it also increases the propeller's cost and complexity. Each additional blade adds surface area, increasing resistance and demanding more power from the engine to rotate effectively.

The hub diameter, 'DH', is the diameter of the central portion of the propeller to which the blades are attached. A larger hub diameter can increase the propeller's structural integrity, but it also increases drag. Consequently, the hub diameter is minimized while ensuring it's sufficiently robust to withstand the stresses induced by propeller operation.

The blade area is where things become more nuanced. The developed area (AD) is the actual surface area of all the blades, as if they were flattened out. The projected area (AP) is the area of the blade outlines projected onto a plane perpendicular to the propeller axis. Lastly, the expanded area (AE) is the area enclosed by the blade outline on a radial section of the propeller. From those three we can find the Blade Area Ratio (BAR), defined as the ratio of the projected blade area to the disc area (the area of the circle swept by the propeller). A higher BAR provides more blade surface for thrust generation, but it also increases drag and the potential for cavitation. It is a critical design parameter that directly impacts propeller performance and efficiency.

The skew angle is the angle of the blade tip relative to a line perpendicular to the propeller axis. Skewing the blades can reduce noise and vibration by distributing the force impulses over a longer period. It also helps to delay the onset of cavitation, which is beneficial for high-speed vessels. However, excessive skew can reduce propeller efficiency and complicate manufacturing.

The rake angle is the angle of the blade's longitudinal axis with respect to a plane perpendicular to the propeller axis. Rake can be positive (blade tips angled aft) or negative (blade tips angled forward). Positive rake can improve propeller efficiency and reduce the bending stresses on the blades, while negative rake can provide better clearance from the hull. The choice of rake angle is typically based on the specific characteristics of the vessel and its operating environment.

The chord length is the distance from the leading edge to the trailing edge of the blade at a specific radial location. It varies along the blade's span, typically being greater near the hub and decreasing towards the tip. The chord length distribution is carefully designed to optimize the lift-to-drag ratio of the blade sections.

The thickness of the blade also changes along the blade length from the root to the tip. Blade thickness distribution is crucial for structural integrity and cavitation performance. Thicker sections provide greater strength but increase drag, whereas thinner sections reduce drag but are more susceptible to cavitation and fatigue.

Finally, cavitation is a phenomenon that must be mentioned. It is the formation of vapor bubbles in regions of low pressure on the propeller blades. These bubbles collapse violently as they move into regions of higher pressure, generating noise, vibration, and erosion damage to the propeller. Careful propeller design, including appropriate blade geometry and operating conditions, is essential to minimize or avoid cavitation.

3.2 Blade Element Theory

Blade Element Theory, often abbreviated as BET, represents a cornerstone in understanding and predicting the performance of marine propellers. Unlike simpler momentum theory, which treats the propeller as a whole and considers overall changes in momentum imparted to the fluid, BET dissects the propeller blade into numerous independent elements. Each element, essentially a small airfoil section, is analyzed individually based on its local geometry and operating conditions. This nuanced approach provides a far more detailed and accurate prediction of propeller thrust, torque, and efficiency, accounting for variations in flow conditions along the blade span. Consequently, understanding the assumptions, limitations, and applications of BET is crucial for anyone involved in propeller design, analysis, or performance evaluation.

The underlying principle behind Blade Element Theory lies in treating each small section of the propeller blade as a rotating airfoil. As the propeller spins, each element experiences a relative inflow velocity resulting from the combination of the vessel's forward speed and the rotational speed of the propeller. This relative inflow velocity, along with the element's angle of attack, determines the lift and drag forces acting on that particular section of the blade. The lift force, primarily acting perpendicular to the relative flow, contributes to the thrust generated by the propeller, while the drag force opposes the motion and contributes to the torque required to rotate the propeller. By integrating the contributions of all blade elements along the entire blade span, one can estimate the total thrust and torque produced by the propeller. Furthermore, the ratio of thrust to torque provides a measure of the propeller's propulsive efficiency, indicating how effectively the propeller converts the engine's power into useful thrust. The beauty of BET is its ability to handle complex blade geometries and varying inflow conditions, making it a versatile tool for propeller analysis.

To apply Blade Element Theory effectively, several assumptions are typically made to simplify the calculations. One crucial assumption is that the flow is two-dimensional, meaning that the flow conditions at each blade element are independent of adjacent elements. This assumption neglects any spanwise flow effects, which can occur due to pressure differences between different sections of the blade. Another key assumption is that the airfoil characteristics, such as lift and drag coefficients, are known for each blade element. These coefficients are usually obtained from experimental data or computational fluid dynamics (CFD) simulations of airfoil sections. It is also commonly assumed that the induced velocities, representing the influence of the propeller wake on the inflow velocity, are uniform across the blade span. In reality, the induced velocities vary along the span, particularly near the blade tips. While these assumptions simplify the analysis, they also introduce potential errors, especially when dealing with highly loaded propellers or complex flow conditions. Ignoring these effects can lead to inaccuracies in thrust and torque predictions.

The actual application of Blade Element Theory involves a series of iterative calculations. First, the propeller blade is divided into a number of discrete elements along its span. For each element, the relative inflow velocity and angle of attack are determined based on the vessel's speed, the propeller's rotational speed, and an initial estimate of the induced velocities. Using the airfoil characteristics, the lift and drag forces acting on each element are then calculated. These forces are resolved into thrust and torque components, which are then integrated along the blade span to obtain the total thrust and torque of the propeller. Next, the calculated thrust and torque are used to update the estimate of the induced velocities. This process is repeated iteratively until the induced velocities and the resulting thrust and torque converge to a stable solution. The convergence usually indicates that the solution is sufficiently accurate. This iterative process can be computationally intensive, especially for propellers with complex geometries or operating conditions. Therefore, computer programs are commonly used to automate the calculations and improve the efficiency of the analysis.

While Blade Element Theory is a powerful tool, it has several limitations that must be considered. As mentioned earlier, the assumption of two-dimensional flow neglects spanwise flow effects, which can be significant for highly loaded propellers or propellers with complex blade geometries. The accuracy of BET also depends heavily on the availability of reliable airfoil characteristics for each blade element. In many cases, experimental data or CFD simulations may not be available for the specific airfoil sections used in the propeller design, leading to uncertainties in the results. Furthermore, BET does not explicitly account for cavitation, a phenomenon that can significantly reduce propeller performance and cause damage. Advanced modifications to the theory may incorporate cavitation models, but these models often rely on empirical correlations and may not be universally accurate. Finally, BET is primarily a steady-state analysis, meaning that it does not account for unsteady effects such as propeller-hull interactions or fluctuations in inflow velocity. These limitations can affect the accuracy of the predictions, especially in complex operating conditions.

Despite its limitations, Blade Element Theory remains a valuable tool for propeller design and analysis. It provides a relatively simple and computationally efficient way to estimate propeller performance and to investigate the effects of various design parameters. It allows designers to rapidly iterate through different propeller geometries and operating conditions to optimize the propeller for specific applications. The detailed information about the flow conditions and forces acting on each blade element can be used to identify areas of high stress or cavitation risk, allowing for targeted design modifications to improve propeller durability and reliability. Furthermore, BET can be used to generate performance curves that show the propeller's thrust, torque, and efficiency as a function of speed and rotational speed. These curves are essential for selecting the appropriate propeller for a given vessel and operating profile.

In conclusion, Blade Element Theory provides a fundamental framework for understanding and predicting the performance of marine propellers. By dissecting the propeller blade into numerous independent elements and analyzing the flow conditions at each element, BET offers a detailed and accurate assessment of propeller thrust, torque, and efficiency. While it has limitations, particularly in accounting for spanwise flow effects and cavitation, BET remains a valuable tool for propeller design and analysis. Its relative simplicity and computational efficiency make it a practical choice for investigating the effects of various design parameters and optimizing propeller performance for specific applications. The insights gained from BET can be used to improve propeller efficiency, reduce fuel consumption, and enhance the overall performance of marine craft. The evolution of this tool is essential for the progress of naval engineering.

3.3 Cavitation Phenomenon

Cavitation is, in its essence, the formation and subsequent collapse of vapor bubbles in a liquid. This phenomenon, while seemingly simple, poses a significant threat to marine propellers and other underwater machinery. Understanding its causes, effects, and mitigation strategies is vital for anyone involved in the design, operation, or maintenance of marine craft. It's more than just bubbles; it's a complex interplay of fluid dynamics and material science.

The root cause of cavitation lies in pressure differentials within the liquid. As a propeller rotates, it creates areas of both high and low pressure around its blades. When the pressure in a localized area drops below the vapor pressure of the water at that temperature, the water begins to boil, forming vapor bubbles. This is similar to what happens when water boils in a pot, but instead of heat, it's the pressure reduction that triggers the phase change. The vapor pressure is dependent on temperature, so warmer water is more susceptible to cavitation. As the propeller blade continues to rotate, these bubbles are carried into regions of higher pressure. The rapid increase in pressure forces the bubbles to collapse violently. This collapse is not a gentle process; it's an implosion, generating intense shock waves and localized micro-jets of liquid.

The effects of cavitation can be devastating to propeller blades. The repeated implosions of vapor bubbles near the blade surface cause a phenomenon known as cavitation erosion. Each implosion acts like a tiny hammer blow, creating stress concentrations on the metal surface. Over time, these repeated stresses lead to fatigue and the removal of material, creating pits and craters on the blade surface. This erosion not only degrades the propeller's performance but also weakens its structural integrity, potentially leading to catastrophic failure. Beyond erosion, cavitation generates noise and vibration. The imploding bubbles create a distinct hissing or crackling sound, often described as sounding like gravel being pumped through the system. This noise can be a nuisance to passengers and crew and can also interfere with sonar systems. The vibrations induced by cavitation can also damage other machinery connected to the propeller shaft, accelerating wear and tear on bearings and seals. Cavitation reduces propeller efficiency. The presence of vapor bubbles disrupts the smooth flow of water around the propeller blades, reducing the thrust generated and increasing drag. This translates to lower vessel speed and increased fuel consumption. It's not a linear relationship either; the severity of these effects increases exponentially as cavitation becomes more pronounced.

Several factors influence the onset and severity of cavitation. Propeller design is critical. A poorly designed propeller with sharp edges or abrupt changes in blade profile is more likely to generate regions of low pressure, promoting cavitation. The operating speed of the propeller is another significant factor. Higher speeds generally lead to lower pressures and increased cavitation risk. The depth of the propeller below the water surface, also known as submergence, plays a crucial role. Greater submergence increases the ambient pressure, making it less likely for the pressure to drop below the vapor pressure. Water temperature affects vapor pressure, where warmer water cavitates easier. Finally, the water quality matters. The presence of dissolved gases or suspended particles in the water can act as nucleation sites, promoting the formation of vapor bubbles and accelerating cavitation.

Fortunately, several strategies can be employed to mitigate cavitation. Optimizing propeller design is paramount. This involves carefully selecting the blade shape, pitch distribution, and blade area ratio to minimize pressure variations and ensure smooth flow. Computational Fluid Dynamics (CFD) simulations can be used to analyze propeller designs and identify areas prone to cavitation. Increasing propeller submergence is another effective approach. By positioning the propeller deeper in the water, the ambient pressure is increased, reducing the likelihood of cavitation. This may involve modifying the hull design or using a larger propeller diameter. Reducing propeller speed can also help to reduce cavitation. While this may impact vessel speed, it can significantly extend the life of the propeller. Using cavitation-resistant materials is a common practice. Materials such as stainless steel, nickel-aluminum bronze, and other alloys are more resistant to cavitation erosion than conventional steel. Applying protective coatings to the propeller blades can provide an additional layer of defense against cavitation damage. These coatings act as a sacrificial layer, absorbing the impact of the collapsing bubbles and preventing damage to the underlying metal. Regular propeller maintenance is essential. Inspecting the blades for signs of cavitation erosion and repairing any damage promptly can prevent further degradation and extend the life of the propeller. This may involve welding, grinding, or polishing the blades to restore their original shape and surface finish.

Understanding the different types of cavitation is also important for effective mitigation. Bubble cavitation, also known as traveling cavitation, involves the formation and collapse of individual bubbles. Sheet cavitation occurs when a large sheet of vapor forms on the blade surface. Vortical cavitation arises from the formation of vapor bubbles within the vortices shed from the propeller tips. Each type of cavitation has its own characteristics and requires different mitigation strategies. Understanding these nuances allows for more targeted and effective interventions.

In summary, cavitation is a complex phenomenon that can have significant consequences for marine propulsion systems. By understanding the causes, effects, and mitigation strategies, we can minimize its impact and ensure the reliable and efficient operation of marine craft. Designing robust propellers, implementing smart operational practices, and conducting routine maintenance are essential to stay ahead of this damaging process. As technology advances, we can expect to see even more sophisticated methods for predicting and preventing cavitation, further enhancing the performance and longevity of marine vessels. This is an ongoing battle, but with knowledge and proactive measures, we can keep the bubbles at bay.

3.4 Propeller Performance Prediction

Predicting the performance of a propeller is a critical aspect of marine craft design and operation, influencing efficiency, speed, and overall vessel performance. The accuracy of these predictions directly impacts design decisions, ensuring that the chosen propeller matches the vessel's requirements and operating conditions. These are obtained through a combination of theoretical models, experimental data, and computational simulations.

Blade Element Momentum Theory (BEMT) stands as a cornerstone in the prediction process. It merges Blade Element Theory (BET), which analyzes the forces on individual blade sections, with Momentum Theory, which considers the overall momentum change in the fluid passing through the propeller disc. BEMT allows us to calculate thrust, torque, and efficiency by integrating the forces along the propeller blades. This iterative process involves determining the inflow angle and induced velocities at each blade element, ensuring that the forces and moments are balanced. The assumptions inherent in BEMT, such as uniform inflow and neglecting radial flow components, must be carefully considered, as they can limit the accuracy of the predictions, especially for heavily loaded propellers. Nevertheless, BEMT provides a valuable starting point for propeller performance analysis and design refinement.

Beyond BEMT, Computational Fluid Dynamics (CFD) offers a more sophisticated approach. CFD simulations solve the Navier-Stokes equations, providing detailed flow field information around the propeller blades and the hull. This enables a more accurate prediction of propeller performance, including the effects of blade geometry, cavitation, and interaction with the hull. To achieve reliable results, careful attention must be paid to grid resolution, turbulence modeling, and boundary conditions. Validation of CFD results with experimental data is essential to ensure the accuracy of the simulations. Although computationally intensive, CFD offers valuable insights into propeller behavior, particularly for complex geometries and operating conditions where BEMT falls short.

Scale model testing, often conducted in towing tanks or cavitation tunnels, provides experimental data for validating theoretical predictions and CFD simulations. Model propellers are tested under controlled conditions, and measurements are taken for thrust, torque, and pressure distribution. These tests can replicate various operating conditions, including different speeds, wake profiles, and cavitation regimes. The challenge lies in scaling the results from the model to the full-scale propeller, accounting for

Reynolds number effects and other scaling factors. Careful consideration must be given to instrumentation accuracy, test setup, and data processing to obtain reliable experimental data. Model testing remains an indispensable tool for propeller performance prediction, particularly for novel designs or complex operating scenarios.

Open water diagrams, typically derived from experimental data or theoretical calculations, provide a graphical representation of propeller performance characteristics. These diagrams plot non-dimensional coefficients such as thrust coefficient (KT), torque coefficient (KQ), and advance coefficient (J). They enable designers to quickly estimate propeller performance for a range of operating conditions. Selecting the appropriate open water diagram for a given propeller geometry and operating regime is crucial for accurate predictions. Limitations in open water diagrams arise from their reliance on specific propeller families and idealized operating conditions. However, they remain a practical tool for preliminary propeller selection and performance estimation.

It is not possible to forget the importance of wake adaptation, an often overlooked aspect. The performance of a propeller is significantly influenced by the wake field it operates in, which is the flow velocity distribution entering the propeller disc. The wake field can be non-uniform due to the hull geometry and appendages. Accurate prediction of the wake field is essential for propeller performance prediction. This can be achieved through wake surveys, CFD simulations of the hull, or empirical methods. Designing the propeller to be adapted to the wake field can improve its efficiency and reduce noise and vibration. This involves optimizing the blade geometry and pitch distribution to account for the non-uniform inflow.

Finally, cavitation represents a critical concern in propeller design. Cavitation occurs when the pressure on the propeller blades drops below the vapor pressure of the water, leading to the formation of vapor bubbles. These bubbles collapse violently, causing noise, vibration, and erosion of the propeller blades. Predicting cavitation requires detailed knowledge of the pressure distribution on the propeller blades and the operating conditions. This can be achieved through CFD simulations, model testing in cavitation tunnels, or empirical methods. Cavitation performance is typically assessed using cavitation number, which represents the ratio of the pressure difference to the dynamic pressure. Propeller designs must aim to minimize cavitation to ensure reliable and efficient operation.

3.5 Advanced Propeller Concepts

Delving into advanced propeller concepts necessitates a shift from basic principles to more nuanced strategies for enhancing performance. We're not just talking about bigger blades or faster rotations; it's about understanding the intricate interplay of hydrodynamic forces and optimizing the propeller to work in harmony with the hull and the specific operational demands of the vessel. This involves considering unconventional designs, materials, and operational strategies to achieve superior efficiency, reduced noise, and minimized environmental impact.

One of the key areas of advancement lies in understanding and mitigating cavitation. While Chapter 3 touched upon this phenomenon, advanced propeller designs actively seek to delay or eliminate cavitation onset. This often involves complex blade geometries, such as skewed or swept blades, which alter the pressure distribution on the blade surface, effectively smoothing out pressure peaks that trigger cavitation. Skewed propellers, where the blade tips are angled backward relative to the direction of rotation, are particularly effective in reducing noise and vibration associated with cavitation. Swept blades, similar to those found on aircraft wings, further refine this pressure distribution, allowing for higher speeds and greater efficiency before cavitation becomes a significant issue. These designs demand sophisticated computational fluid dynamics (CFD) analysis to predict their performance accurately, requiring a deeper understanding of the fluid flow around the propeller.

Furthermore, the application of computational methods extends beyond just blade geometry. We can now optimize the entire propeller-hull interaction. Wake-adapted propellers are designed specifically to operate within the non-uniform wake field behind a ship's hull. The wake is the disturbed flow of water created by the hull as it moves through the water. This flow is often turbulent and uneven, presenting a challenge for a propeller to operate efficiently. By carefully tailoring the propeller blade shape and pitch distribution to match the wake profile, it's possible to achieve significant improvements in propulsive efficiency. Imagine it like fitting a puzzle piece perfectly into its space; the wake-adapted propeller is custom-fitted to the flow conditions it will encounter. This requires detailed measurements or simulations of the wake field, followed by iterative design and analysis to achieve the optimal propeller geometry. It's a complex process, but the potential gains in efficiency are substantial, especially for larger vessels operating at high speeds.

Consider the rise of unconventional propeller configurations. Contra-rotating propellers, consisting of two propellers rotating in opposite directions on the same axis, represent a significant departure from traditional designs. The primary advantage is the recovery of swirl energy. A conventional propeller imparts a swirling motion to the water as it accelerates it backward. This swirl represents wasted energy. By placing a second propeller behind the first, rotating in the opposite direction, this swirl energy can be recovered, resulting in a higher overall propulsive efficiency. Contra-rotating propellers are particularly effective at higher speeds and for vessels where fuel efficiency is paramount. However, they also present challenges in terms of mechanical complexity and maintenance, requiring precise alignment and robust bearings to withstand the increased loads.

Another promising avenue is the development and use of advanced materials. Traditional propeller materials, such as nickel-aluminum bronze, have served the industry well for decades. However, new materials offer the potential for increased strength, reduced weight, and improved resistance to corrosion and cavitation erosion. Composite materials, such as carbon fiber reinforced polymers, are gaining traction in propeller design. These materials offer a high strength-to-weight ratio, allowing for the creation of lighter and more efficient propellers. Lighter propellers reduce the overall weight of the propulsion system and can improve the vessel's maneuverability and response time. Furthermore, composite materials can be tailored to specific performance requirements, allowing for the optimization of blade flexibility and damping characteristics. Coatings also play a crucial role. Applying specialized coatings to propeller blades can reduce friction, inhibit fouling, and provide enhanced protection against cavitation erosion. These coatings can be engineered to have specific surface properties, such as hydrophobicity (water-repelling) or self-polishing capabilities, further enhancing propeller performance and longevity.

Think about noise reduction. Propeller noise is not only a nuisance for passengers and crew but also a significant environmental concern, particularly for marine life. Cavitation is a major source of propeller noise, but other factors, such as blade singing (vibration of the blades due to hydrodynamic forces), can also contribute. Advanced propeller designs incorporate features to minimize noise generation. This can include carefully shaped blade tips, optimized blade thickness distributions, and the use of damping materials to reduce vibration. Furthermore, computational aeroacoustics (CAA) techniques are now being used to predict and mitigate propeller noise during the design phase. CAA simulations can identify noise hotspots on the propeller blades and guide the optimization of the geometry to minimize noise radiation. The pursuit of quieter propellers is not just about comfort; it's about protecting the marine environment and ensuring the sustainability of maritime activities.

In the end, the pursuit of advanced propeller concepts is a continuous process of innovation and refinement. It requires a multidisciplinary approach, bringing together expertise in hydrodynamics, materials science, and computational methods. As we push the boundaries of propeller technology, we must also consider the broader implications for vessel design, operational efficiency, and environmental sustainability.

4 Chapter 4: Alternative Propulsion Methods

Stepping away from the familiar sight of propellers churning through water, we find ourselves entering a realm of innovation where alternative propulsion methods reign supreme. This chapter is dedicated to unveiling these systems, particularly waterjets and podded drives (including azimuthing thrusters), peeling back their operational layers and highlighting the specific circumstances where they outshine traditional propeller setups. I remember the first time I saw a waterjet in action; the sheer power and maneuverability were astonishing, quite different from the steady hum of a propeller-driven vessel. This experience cemented my fascination with these alternative technologies, and I'm eager to share this knowledge with you.

Waterjet propulsion systems function on a remarkably straightforward principle: water is drawn into the unit, accelerated to high velocity, and then forcefully expelled through a nozzle, generating thrust. It's akin to a jet engine, but instead of air, it uses water as the working fluid. The key advantage of waterjets lies in their ability to operate efficiently at high speeds, making them a natural fit for fast ferries, patrol boats, and other high-performance vessels. Imagine a sleek, high-speed ferry effortlessly slicing through waves; the chances are high that it's propelled by a waterjet system. This is because, unlike propellers which suffer from cavitation and reduced efficiency at higher speeds, waterjets maintain their performance, delivering superior thrust and maneuverability. Furthermore, waterjets are less susceptible to damage from underwater debris, making them a practical choice for vessels operating in shallow or debris-filled waters. Consider a rescue vessel operating in a harbor; a waterjet would allow it to navigate through cluttered waters with ease, without the risk of damaging a propeller. The design of a waterjet incorporates an inlet duct to guide water into the impeller, which is essentially a high-speed pump. This impeller then accelerates the water and directs it towards the nozzle. Varying the nozzle area allows for thrust control, enabling the vessel to adjust its speed and direction. The steering mechanism often involves deflectors that redirect the water jet stream, providing excellent maneuverability.

However, waterjets are not without their limitations. They tend to be less efficient than propellers at lower speeds, making them unsuitable for vessels that primarily operate at displacement speeds. Their higher initial cost and increased complexity in maintenance can also be a deterrent for some applications. Another consideration is the noise generated by the high-speed water flow, which can be a concern in certain environments. In essence, the decision to use waterjets hinges on a careful evaluation of the vessel's operational profile, speed requirements, and budget considerations. Selecting the right waterjet for a vessel is not as simple as picking one from a catalogue, it involves carefully sizing the waterjet to provide the required thrust at the vessel's operating speed. This involves advanced calculations and computer simulations to ensure optimal performance.

Moving on to podded drives, these innovative propulsion units represent a significant departure from traditional shaft lines and rudders. A podded drive essentially integrates an electric motor directly into a streamlined housing (the "pod") that is mounted externally beneath the vessel's hull. The propeller is attached directly to this motor, eliminating the need for long propeller shafts and gears. Azimuthing thrusters, a subset of podded drives, take this concept a step further by allowing the entire pod to rotate 360 degrees, providing unparalleled maneuverability. Think of a tugboat effortlessly spinning in place; the chances are high that it's equipped with azimuthing thrusters. This rotational capability allows for precise control of thrust direction, enabling vessels to move sideways, diagonally, or even rotate on the spot. Podded drives offer several key advantages. Firstly, they improve hydrodynamic efficiency by reducing drag compared to traditional shaft lines and rudders. The streamlined pod design minimizes resistance, leading to improved fuel efficiency and reduced emissions. Secondly, they enhance maneuverability, particularly for vessels operating in confined spaces. The ability to precisely control thrust direction allows for precise docking, harbor maneuvering, and dynamic positioning. Thirdly, they simplify ship design by eliminating the need for a separate rudder and steering gear. The integrated design reduces space requirements and allows for greater flexibility in hull layout.

Consider a large cruise ship docking in a crowded port. The captain can leverage the precise control offered by podded drives to smoothly maneuver the ship into its berth, even in challenging wind and current conditions. A critical aspect of podded drive design is the electric motor. These motors are typically synchronous or induction motors, powered by onboard generators. The use of electric propulsion allows for greater flexibility in power distribution and control, enabling efficient operation across a wide range of speeds and loads. Moreover, podded drives often incorporate advanced control systems that optimize propeller pitch and thrust direction to maximize efficiency and minimize noise.

However, podded drives also come with certain drawbacks. Their initial cost is generally higher than traditional propulsion systems, and their maintenance can be more complex. The external location of the pod makes it vulnerable to damage from collisions or grounding, and repairs can be challenging and costly. Furthermore, the electrical power requirements of podded drives can be significant, requiring substantial generator capacity. The complexity of the electric drive components also necessitates highly skilled technicians for maintenance and repair. When considering podded drives, it's essential to factor in the increased maintenance costs and the need for specialized training for the crew. Despite these challenges, the benefits of podded drives in terms of efficiency and maneuverability make them an attractive option for a wide range of vessels, including cruise ships, ferries, icebreakers, and offshore support vessels. Their adoption

is steadily growing, driven by the increasing demand for cleaner, more efficient, and more maneuverable ships. The increasing environmental regulations and rising fuel costs are further incentivizing the adoption of these innovative technologies.
In conclusion, the choice between waterjets, podded drives, and traditional propellers depends heavily on the specific application. Waterjets excel at high speeds and in shallow water, while podded drives offer superior maneuverability and efficiency for a wider range of vessel types. As technology continues to advance, we can expect to see even more innovative propulsion methods emerge, further pushing the boundaries of marine craft efficiency and performance. I believe that understanding these alternative technologies is crucial for anyone involved in the design, operation, or management of marine vessels. Embracing these innovations will not only improve the performance of our ships but also contribute to a more sustainable and environmentally friendly maritime industry.

4.1 Waterjet Propulsion Systems

Waterjet Propulsion Systems
The allure of waterjet propulsion lies in its unique application: vessels craving high speeds and agile maneuverability. Unlike propellers which churn through the water, waterjets operate within a contained system, drawing water in through an intake, accelerating it via an impeller, and expelling it through a nozzle to generate thrust. This distinction provides advantages in shallow waters, reducing the risk of propeller damage, and offer-

ing a quieter operation compared to traditional systems due to the enclosed mechanism. Consider, for instance, high-speed ferries gliding through bustling harbors or military crafts navigating complex coastal terrains; these are prime examples of where waterjets showcase their superiority. But let's delve deeper than just observing its use.

The core of a waterjet system lies in its components, each playing a crucial role in the overall performance. The intake, strategically positioned on the hull, is designed to minimize turbulence and ensure a smooth flow of water into the system. The impeller, akin to a propeller inside a duct, accelerates the water. Impeller design varies considerably, from axial-flow to mixed-flow configurations, each optimized for specific speed ranges and vessel types. The stator vanes, located downstream of the impeller, straighten the swirling flow, converting rotational energy into axial thrust. And finally, the nozzle, a converging duct, increases the water velocity, maximizing thrust output. Understanding the interplay between these elements is paramount to appreciating the system's function as a whole, and how seemingly small design choices drastically alter performance.

One of the most crucial aspects of waterjet propulsion is its influence on vessel maneuverability. By manipulating the direction of the exiting water jet, either through vectoring nozzles or reversing buckets, exceptional control can be achieved. Vectoring nozzles allow for precise steering, enabling rapid changes in direction and tight turning circles. Reversing buckets, deflectors placed behind the nozzle, redirect the water flow forward, providing effective braking and reversing capabilities. Think of a patrol boat executing a sharp turn at high speed or a tugboat deftly maneuvering in a crowded harbor. These maneuvers are not possible without the responsiveness provided by such systems.

Diving into the specifics of design considerations, one finds that impeller selection is a critical decision. Axial-flow impellers are generally preferred for high-speed applications due to their ability to generate high flow rates. Mixed-flow impellers offer a compromise between flow rate and pressure, making them suitable for a wider range of vessel speeds. The nozzle design also plays a significant role, influencing the jet velocity and thrust. Convergent nozzles increase jet velocity but can also increase energy losses due to friction. Divergent nozzles can reduce losses but may also reduce thrust. Careful consideration must be given to matching the impeller and nozzle characteristics to the specific vessel requirements.

Efficiency considerations within waterjet systems are complex and multi-faceted. While waterjets offer advantages in maneuverability and shallow water operation, they often suffer from lower propulsive efficiency compared to propellers, particularly at lower speeds. This is primarily due to losses associated with the intake, impeller, and nozzle. To mitigate these losses, designers focus on optimizing the intake geometry to minimize turbulence, using high-efficiency impellers with carefully designed blade profiles, and employing nozzle designs that minimize frictional losses. The jet velocity also affects efficiency; higher jet velocities result in higher thrust but also higher energy losses. Finding the optimal balance between jet velocity and thrust is crucial for maximizing efficiency.

Comparing waterjets to traditional propellers reveals clear distinctions in performance characteristics. Propellers generally exhibit higher efficiency at lower speeds, making them suitable for displacement hulls and vessels that spend a significant amount of time cruising at moderate speeds. Waterjets, on the other hand, excel at high speeds, where propeller efficiency tends to decrease due to cavitation and other hydrodynamic effects. Waterjets also offer superior maneuverability, particularly in shallow water, where propellers are prone to damage. The choice between a waterjet and a propeller depends heavily on the specific vessel application and operating profile.

When considering specific applications, several examples highlight the benefits of waterjet propulsion. High-speed ferries, often operating in congested waterways, benefit from the maneuverability and speed capabilities of waterjets. Military vessels, such as patrol boats and special forces craft, utilize waterjets for their speed, shallow water operation, and reduced noise signature. Yachts and pleasure craft often employ waterjets for their smooth ride, quiet operation, and ease of handling. Each application showcases unique advantages that waterjets bring to the table, fulfilling specific operational needs.

The maintenance and operational aspects of waterjet systems also merit attention. Regular inspection of the intake, impeller, and nozzle is essential to ensure optimal performance and prevent damage. Impeller wear and erosion can reduce efficiency and increase noise levels. Nozzle damage can affect thrust and maneuverability. Proper lubrication of bearings and seals is crucial for preventing premature wear. Regular maintenance intervals should be established based on the operating environment and vessel usage. A preventative maintenance program can significantly extend the lifespan of the system and minimize downtime.

Innovations continue to push the boundaries of waterjet technology. Variable geometry nozzles, which can adjust the jet exit area to optimize performance across a range of speeds, are being developed. Advanced impeller designs, incorporating computational fluid dynamics (CFD) analysis, are improving efficiency and reducing cavitation. The integration of advanced control systems, including dynamic positioning and autopilot, is enhancing maneuverability and ease of operation. These advancements promise to further improve the performance and versatility of waterjet propulsion systems.

Finally, considering the environmental impact of waterjet systems, several factors come into play. While waterjets themselves do not directly emit pollutants, the fuel consumption of the vessel does contribute to emissions. Reducing fuel consumption through improved efficiency and the use of alternative fuels can mitigate these impacts. The noise generated by waterjets can also affect marine life, although it is generally lower than that of propellers. Careful design and operation can minimize noise levels and reduce the environmental footprint of waterjet propelled vessels. As technology advances, expect further innovation in reducing the environmental impact of these dynamic propulsion systems.

4.2 Podded Drives and Azimuthing Thrusters

Podded drives and azimuthing thrusters represent a significant departure from traditional marine propulsion systems, offering enhanced maneuverability and efficiency for a variety of vessel types. These systems, essentially self-contained propulsion units mounted externally to the hull, have revolutionized how ships are designed and operated, particularly in applications requiring precise control and station-keeping. The core concept involves integrating the propeller directly with an electric motor within a streamlined pod, which can then be rotated to direct thrust in any direction, eliminating the need for conventional rudders. This integrated design offers several advantages, but also presents unique challenges in terms of design, operation, and maintenance.

The fundamental principle behind podded drives lies in their ability to generate thrust in a 360-degree range, providing exceptional maneuverability. This is achieved through the azimuthing capability, where the entire pod rotates around a vertical axis. This allows vessels to move laterally, pivot on their axis, and execute complex maneuvers with a level of precision unattainable with traditional rudder-propeller systems. The integration of the electric motor directly within the pod eliminates the need for long propeller shafts and complex gearing arrangements, resulting in improved mechanical efficiency and reduced vibration. Furthermore, the compact design frees up valuable internal space within the hull, which can be utilized for cargo, passenger accommodation, or other operational needs. Different configurations exist, with pulling and pushing arrangements. The pulling propeller, positioned at the front of the pod, tends to experience less disturbed water flow, potentially leading to higher propulsive efficiency, but can be more susceptible to cavitation if not carefully designed. Pushing propellers, located at the rear, benefit from a cleaner inflow environment due to the pod body shielding the propeller.

Azimuthing thrusters, while similar in function to podded drives, often differ in their specific design and application. Typically, azimuthing thrusters are driven by diesel engines or electric motors located within the hull, with power transmitted to the thruster unit via shafts and gears. While they offer similar maneuverability benefits as podded drives, they may not achieve the same level of mechanical efficiency due to the power transmission losses. These thrusters are commonly used in dynamic positioning (DP) systems, where precise control of vessel position and heading is crucial. DP systems rely on multiple thrusters, often including azimuthing thrusters, to counteract environmental forces such as wind, waves, and currents, maintaining the vessel's position within a defined tolerance. Azimuthing thrusters are also frequently employed in tugboats, ferries, and offshore support vessels, where their maneuverability enhances operational capabilities in confined spaces and challenging environments. The selection between podded drives and azimuthing thrusters depends on a variety of factors, including the vessel type, operational requirements, and economic considerations.

Effective utilization of podded drives and azimuthing thrusters hinges on a comprehensive understanding of their operational characteristics and control systems. Sophisticated control algorithms are essential to manage the thrust and direction of multiple thrusters, coordinating their actions to achieve desired vessel movements. These algorithms take into account factors such as vessel speed, heading, and environmental conditions, constantly adjusting thruster outputs to maintain stability and control. Operator training is also paramount, as maneuvering vessels equipped with these systems requires a different skillset compared to traditional rudder-propeller vessels. Familiarization with the control interfaces, system limitations, and emergency procedures is crucial for safe and efficient operation. Additionally, regular maintenance and inspection are vital to ensure the reliability and longevity of these complex systems. This includes monitoring motor performance, gear lubrication, and seal integrity, as well as conducting periodic underwater inspections to assess the condition of the pod or thruster units. Special attention should be paid to corrosion protection, as these components are constantly exposed to seawater.

However, the adoption of podded drives and azimuthing thrusters also presents certain challenges. The initial investment cost is typically higher compared to conventional propulsion systems, requiring a careful cost-benefit analysis to justify the expenditure. The complexity of these systems also necessitates specialized maintenance and repair expertise, which may not be readily available in all locations. Concerns about the hydrodynamic efficiency of podded drives, particularly at higher speeds, have also been raised. The presence of the pod body can disrupt water flow around the propeller, potentially increasing drag and reducing overall efficiency. However, ongoing research and development efforts are focused on optimizing pod designs to minimize these effects. Another consideration is the potential for increased noise and vibration levels, which can impact crew comfort and marine life. Noise mitigation strategies, such as vibration damping and acoustic insulation, are often employed to address these issues. Despite these challenges, the benefits of enhanced maneuverability, improved efficiency, and space savings have made podded drives and azimuthing thrusters an increasingly popular choice for a wide range of marine applications.

Looking ahead, further advancements in podded drive and azimuthing thruster technology are anticipated. The integration of advanced sensor technology and data analytics will enable more precise control and optimize energy consumption. The use of composite materials in pod construction will reduce weight and improve hydrodynamic performance. Furthermore, the development of electric propulsion systems powered by renewable energy sources, such as fuel cells and batteries, will further enhance the sustainability of these propulsion methods. As environmental regulations become increasingly stringent, podded drives and azimuthing thrusters are poised to play an even greater role in shaping the future of marine propulsion.

5 Chapter 5: Energy Efficiency Strategies

Energy efficiency isn't just a buzzword; it's the cornerstone of modern maritime operations, demanding a multifaceted approach that touches every aspect of a vessel's design and operation. To truly grasp the potential for optimization, we must dive deep into strategies that extend beyond simply using more efficient engines. These strategies can be organized to hull form optimization, friction reduction, and waste heat recovery.

Optimizing the hull form is probably the most impactful step one can take to slash fuel consumption. A well-designed hull minimizes resistance, allowing the vessel to glide through the water with less power. Naval architects meticulously shape the hull using computational fluid dynamics (CFD) simulations and tank testing to identify areas of high resistance and refine the design accordingly. Bulbous bows, for example, strategically placed at the bow of the ship, can reduce wave-making resistance by creating a wave that cancels out the bow wave. Similarly, stern wedges can improve the flow of water around the stern, reducing drag and increasing propulsive efficiency. However, hull form optimization isn't a one-size-fits-all solution. The ideal hull shape depends on the vessel's intended speed, operating conditions, and cargo capacity. A fast container ship will have a drastically different hull form than a slow-moving bulk carrier. The process can be expensive, involving sophisticated modeling and testing, but the long-term fuel savings and reduced emissions make it a worthwhile investment. Furthermore, hull design is not static; regular inspections and maintenance are crucial to identify and address any damage or fouling that can increase resistance. Attention to detail during the design phase, coupled with ongoing maintenance, is crucial for maximizing fuel efficiency.

Friction reduction techniques represent another critical area for enhancing energy efficiency. As a vessel moves through the water, friction between the hull and the water creates a significant amount of drag. To combat this, various strategies can be employed, ranging from advanced coatings to innovative air lubrication systems. Antifouling coatings are essential for preventing the growth of marine organisms on the hull. These organisms, such as barnacles and algae, increase surface roughness and significantly increase frictional resistance. Traditional antifouling coatings used toxic substances, but modern coatings are increasingly using more environmentally friendly alternatives. Furthermore, regular hull cleaning is vital to remove any fouling that does accumulate, even with the best antifouling coatings. Air lubrication systems, a relatively newer technology, inject air bubbles beneath the hull, creating a layer of air between the hull and the water. This reduces the wetted surface area and, consequently, the frictional resistance. While air lubrication systems require additional energy to compress and inject the air, the overall energy savings can be substantial, particularly for large vessels operating at high speeds. Choosing the right combination of friction reduction techniques depends on factors like vessel type, operating speed, and environmental regulations. Regular monitoring of hull condition and performance is essential to ensure that these techniques are functioning optimally. These systems are an investment in efficiency.

Waste heat recovery (WHR) systems offer a unique opportunity to reclaim otherwise lost energy and convert it into useful work. Marine engines, like any combustion engine, reject a significant portion of their energy as heat in the exhaust gas and cooling water. WHR systems capture this waste heat and use it to generate electricity, produce steam, or preheat fuel or water, reducing the overall fuel consumption of the vessel. Steam Rankine Cycle (SRC) systems are commonly used for WHR. These systems use the waste heat to boil water, creating steam that drives a turbine connected to a generator. Organic Rankine Cycle (ORC) systems are similar to SRC systems but use an organic fluid with a lower boiling point, making them suitable for lower-temperature waste heat sources. Thermoelectric generators (TEGs) directly convert heat into electricity using the Seebeck effect. While TEGs have lower efficiency than SRC or ORC systems, they are compact and require minimal maintenance. The economic viability of WHR systems depends on the amount and temperature of the available waste heat, as well as the cost of the equipment. Larger vessels with high engine loads are typically better candidates for WHR systems. Moreover, the integration of WHR systems requires careful design and optimization to ensure that they operate efficiently and reliably. Despite the complexity, WHR technologies offer a substantial opportunity to improve the energy efficiency of marine vessels and reduce their environmental footprint.

5.1 Hull Form Optimization

Hull form optimization is more than just making a ship look sleek; it's about sculpting a vessel to glide through water with minimal resistance. For a beginner, the sheer number of variables can feel overwhelming, but understanding the fundamental principles is the key to unlocking significant efficiency gains. We're talking about reducing fuel consumption, increasing speed, and even improving stability - all from the shape of the hull. It starts with grasping the different types of resistance a ship encounters.
Think of it this way: imagine pushing a flat board through water versus pushing a streamlined fish. The flat board meets significant resistance, creating a large wake and requiring considerable force. The fish, on the other hand, slips through more easily, creating a smaller wake and needing less energy. This analogy highlights the core concept of hull form optimization: shaping the hull to minimize resistance. Two primary types of resistance are at play: frictional resistance and wave-making resistance. Frictional resistance is the drag caused by the water flowing along the hull's surface. Wave-making resistance arises from the energy spent creating waves as the ship moves. A well-optimized hull form minimizes both of these. It is a delicate balancing act.
Now, let's get specific. One of the first parameters to consider

is the Length-to-Beam Ratio (L/B). A longer, narrower hull (high L/B ratio) generally experiences lower wave-making resistance, especially at higher speeds. However, it can increase frictional resistance due to the larger wetted surface area. Conversely, a shorter, wider hull (low L/B ratio) offers greater stability and maneuverability, but at the cost of increased wave-making resistance. For beginners, a good starting point is to research the typical L/B ratios for the specific type of vessel you're considering. Cargo ships, for example, tend to have higher L/B ratios than tugboats, reflecting their different operational requirements. The block coefficient (Cb) is another vital parameter. This dimensionless number represents the ratio of the underwater volume of the hull to the volume of a rectangular block with the same length, beam, and draft. A lower Cb indicates a finer, more streamlined hull, typically leading to lower resistance at higher speeds. However, a lower Cb also means less cargo-carrying capacity. Therefore, the optimal Cb depends on the vessel's primary function and operating speed. Remember, there's no one-size-fits-all solution. You need to carefully consider the trade-offs between speed, capacity, and stability. These considerations become especially important when dealing with unique vessels.

Moving beyond these general parameters, consider the shape of the bow and stern. The bow is the first part of the ship to encounter the water, and its shape significantly influences wave-making resistance. A bulbous bow, a protruding bulb-shaped structure below the waterline, can reduce wave-making resistance by creating a wave that cancels out the bow wave generated by the hull. The design of a bulbous bow is a complex process, dependent on the ship's speed and hull form. The stern, similarly, plays a crucial role in reducing resistance and improving propulsive efficiency. A well-designed stern allows for a smooth flow of water to the propeller, minimizing turbulence and maximizing thrust. Different stern shapes, such as transom sterns and canoe sterns, offer varying advantages depending on the vessel's intended use. Transom sterns are common in many ships, offering a relatively simple and efficient design. Canoe sterns, with their more rounded shape, can reduce wave-making resistance at higher speeds. Choosing the appropriate stern shape is essential for optimizing overall vessel performance. The devil is in the detail.

Alright, so how do you actually optimize a hull form? For beginners, it's best to start with existing designs and then make incremental modifications. Naval architecture software can be incredibly helpful, but it's essential to understand the underlying principles before diving into complex simulations. Computational Fluid Dynamics (CFD) software allows you to simulate the flow of water around a hull, providing valuable insights into resistance and flow patterns. Tank testing, where a scale model of the hull is tested in a towing tank, is another valuable tool for validating designs and measuring resistance. These methods provide data to refine the hull. These tools can be expensive and require specialized knowledge, but even simple models can provide valuable insights. Think about investing in educational software to start.

Let's also discuss coatings. While not strictly part of the hull form itself, the surface finish significantly impacts frictional resistance. Smooth, clean surfaces reduce friction, while rough or fouled surfaces increase it. Anti-fouling coatings are essential for preventing the growth of marine organisms on the hull, which can drastically increase resistance. Regular hull cleaning is also crucial for maintaining optimal performance. Emerging technologies, such as air lubrication systems, can further reduce frictional resistance by creating a layer of air between the hull and the water. These systems, while more complex to implement, can offer significant fuel savings, especially for larger vessels. I have seen the impact.

Finally, keep in mind that hull form optimization is an iterative process. It's not about finding the perfect shape on the first try, but about continuously refining the design based on testing and analysis. As you gain experience, you'll develop a better understanding of the complex interplay between different hull parameters and their impact on vessel performance. Don't be afraid to experiment and learn from your mistakes. Every vessel is unique, and the optimal hull form will depend on its specific requirements and operating conditions. Remember that sustainability is becoming increasingly important. The principles of hull form optimization, when combined with other energy-efficient technologies, can contribute significantly to reducing the environmental impact of marine transportation. Embrace the challenge, and you'll be well on your way to designing more efficient and sustainable vessels.

5.2 Friction Reduction Techniques

Friction, an unavoidable consequence of a vessel's movement through water, significantly impacts fuel consumption and overall efficiency. Managing and minimizing this resistance is paramount for designing energy-efficient marine craft. Several strategies exist to tackle this challenge, each with its own set of advantages, limitations, and application scenarios. Understanding these techniques is essential for naval architects and marine engineers striving to optimize vessel performance.

One of the most direct approaches to reducing friction is through the application of specialized coatings to the hull. These coatings, broadly categorized as either foul-release or ablative, aim to minimize the growth of marine organisms, known as biofouling. Biofouling drastically increases frictional resistance by creating a rougher surface and disrupting the smooth flow of water. Foul-release coatings create a slick, non-stick surface that prevents organisms from adhering in the first place. These coatings often incorporate silicone-based polymers that reduce surface energy, making it difficult for organisms to gain a foothold. Regular cleaning may still be required, but the ease of removal is significantly enhanced. Ablative coatings, on the other hand, contain biocides that are slowly released into the surrounding water, preventing the settlement of organisms. As the vessel moves through the water, a thin layer of the coating erodes, continuously exposing fresh biocide and smoothing the surface. While effective, ablative coatings raise environmental concerns due to the release of biocides into the marine environment, leading to increasingly stringent regulations and the development of more eco-friendly alternatives. Furthermore, the effectiveness of both coating types can be influenced by factors like water temperature, salinity, and the specific types of organisms present in the operating environment. Therefore, selecting the appropriate coating requires careful consideration of the vessel's intended operating profile and environmental impact.

Another promising technique for reducing frictional resistance is air lubrication. This method involves injecting a layer of air between the hull and the water, effectively reducing the wetted surface area and, consequently, the frictional drag. Various implementations of air lubrication exist, ranging from air cavity systems, where a recessed cavity in the hull is filled with air, to air layer systems, where a thin layer of air is distributed across the hull surface through strategically placed air injectors. Air cavity systems generally offer greater drag reduction at higher speeds but can be more complex to implement and may impact stability, while air layer systems are more suitable for lower speeds and offer a more uniform reduction in friction. The effectiveness of air lubrication is highly dependent on factors such as the vessel's speed, hull shape, and the rate of air injection. Careful optimization of these parameters is crucial to achieving significant drag reduction without compromising stability or increasing energy consumption for air compression. Challenges associated with air lubrication include the need for robust air injection systems, the potential for increased noise and vibration, and the risk of air layer instability in rough seas. Despite these challenges, air lubrication holds significant potential for improving the energy efficiency of large vessels, particularly those operating on relatively calm waters.

Beyond coatings and air lubrication, hull form optimization plays a crucial role in minimizing frictional resistance. The shape of the hull directly influences the flow of water around the vessel, and even small changes in hull geometry can have a significant impact on drag. A streamlined hull shape with smooth, gradual curves minimizes flow separation and reduces the formation of eddies, which contribute to frictional resistance. The design process typically involves extensive computational fluid dynamics (CFD) simulations to analyze the flow field around the hull and identify areas where drag can be reduced. Optimizing the hull form often involves trade-offs between different types of resistance, such as frictional resistance and wave-making resistance. For example, a hull shape that minimizes frictional resistance may not be ideal for reducing wave-making resistance at higher speeds. Therefore, the optimal hull form depends on the vessel's intended operating speed and sea conditions. Furthermore, hull form optimization must consider other factors such as stability, cargo capacity, and maneuverability. Advanced hull forms, such as those incorporating bulbous bows or stern wedges, can further reduce resistance and improve overall efficiency.

Maintaining a smooth hull surface is also essential for minimizing friction. Even without biofouling, imperfections in the hull surface, such as scratches, dents, or weld seams, can increase frictional resistance. Regular hull cleaning and maintenance are therefore crucial for preserving a smooth surface and minimizing drag. The use of high-quality paints and coatings can also help to protect the hull surface from corrosion and abrasion. Furthermore, the application of fairing compounds can smooth out irregularities and further reduce friction. The frequency of hull cleaning depends on the vessel's operating environment and the type of coating used. Vessels operating in heavily fouled waters may require more frequent cleaning than those operating in cleaner waters. Regular inspections can help to identify areas where maintenance is needed and prevent the build-up of fouling or damage to the hull surface.

In conclusion, reducing friction is a multifaceted challenge that requires a combination of strategies. From advanced coatings and air lubrication to hull form optimization and regular maintenance, each technique plays a vital role in minimizing drag and improving the energy efficiency of marine craft. The selection of the most appropriate techniques depends on a variety of factors, including the vessel's size, speed, operating environment, and budget. Continuous research and development are leading to new and improved friction reduction technologies, promising further advancements in marine vessel efficiency in the years to come.

5.3 Waste Heat Recovery Systems

Imagine the engine room of a large container ship, a space typically characterized by the relentless roar of massive diesel engines. However, this chapter isn't about generating that power, but about what happens to all the energy that doesn't end up turning the propeller. A staggering amount of energy, a significant portion of the fuel's potential, is lost as heat. This heat radiates from engine surfaces, exits through exhaust stacks, and dissipates within cooling systems. For too long, this waste heat has been treated as an unavoidable byproduct. But with growing pressures to reduce emissions and improve fuel efficiency, innovative Waste Heat Recovery (WHR) systems are becoming increasingly vital.

These systems aren't just theoretical concepts; they're tangible technologies transforming how we think about energy utilization on ships. The basic principle is straightforward: capture the waste heat and convert it into a usable form of energy, typically electricity or steam. The recovered energy can then be used to power auxiliary equipment, such as pumps, lighting, and air conditioning, reducing the load on the main generators and, consequently, fuel consumption. Think of it as a second chance for energy, a way to extract additional value from every liter of fuel burned. The potential benefits are substantial, including reduced fuel costs, lower greenhouse gas emissions, and increased overall operational efficiency. We're not talking about marginal improvements; we're discussing significant savings that can dramatically impact a vessel's operating costs and environmental footprint over its lifespan.

Different WHR technologies cater to different types and quantities of waste heat. Exhaust Gas Boilers (EGBs) are among the most common, typically found on larger vessels with substantial exhaust heat streams. They function much like conventional boilers, using the hot exhaust gases to heat water and produce steam. This steam can then be used to drive a steam turbine, generating electricity. The size and complexity of an EGB system depend on the size of the engine and the amount of recoverable heat. Another technology is the Organic Rankine Cycle (ORC), which uses an organic fluid with a lower boiling point than water as the working fluid. This allows the system to operate efficiently at lower temperatures, making it suitable for recovering heat from jacket water and other lower-temperature sources. ORC systems are generally more compact and require less maintenance than steam turbine systems, making them attractive for smaller vessels or applications where space is limited. Then there are turbochargers, which are inherently energy recovery devices, using exhaust gas energy to compress intake air, thereby improving engine efficiency. Advanced turbocharger designs further optimize this process, extracting even more energy from the exhaust stream. Thermo-electric Generators (TEGs) represent a more novel approach. They directly convert heat into electricity using the Seebeck effect. While TEGs have lower efficiency compared to EGBs or ORC systems, they are solid-state devices with no moving parts, resulting in high reliability and low maintenance requirements. These technologies can be used in different combination.

The effectiveness of a WHR system depends critically on careful design and integration with the existing vessel systems. A detailed heat balance analysis is crucial to accurately quantify the available waste heat at different operating conditions. This analysis informs the selection of the appropriate WHR technology and determines the optimal size and configuration of the system. The integration process must consider the existing power distribution system, the steam network (if applicable), and the cooling water circuits. Careful attention must be paid to avoid any negative impacts on engine performance or reliability. This can involve modifications to the engine control system, the cooling system, and the exhaust system. The return on investment for a WHR system depends on several factors, including the cost of fuel, the operating profile of the vessel, and the capital cost of the system. Vessels operating on long voyages with high engine loads typically benefit the most from WHR systems. However, even vessels with shorter voyages can achieve significant savings with properly designed and optimized systems.

Implementing WHR systems presents both challenges and opportunities for the maritime industry. One of the main barriers is the initial capital cost. WHR systems can be expensive, requiring a significant upfront investment. However, the long-term fuel savings can often offset this cost within a few years. Another challenge is the space and weight requirements. WHR systems can be bulky and heavy, which can be a concern for vessels with limited space. The complexity of integrating WHR systems with existing vessel systems can also be a challenge. However, with proper planning and engineering, these challenges can be overcome. I feel strongly that the opportunities are vast. WHR systems are not just about saving money; they're about creating a more sustainable and environmentally responsible maritime industry. By reducing fuel consumption, we can reduce greenhouse gas emissions and air pollution. By increasing energy efficiency, we can reduce our reliance on fossil fuels.

Looking ahead, the future of WHR in marine applications is

bright. Ongoing research and development efforts are focused on improving the efficiency, reducing the cost, and increasing the reliability of WHR technologies. Advanced materials, such as high-temperature alloys and nano-materials, are being explored to enhance the performance of EGBs, ORC systems, and TEGs. Integrated energy management systems, which combine WHR with other energy efficiency technologies, are also gaining traction. These systems use sophisticated control algorithms to optimize the overall energy performance of the vessel. Furthermore, stricter environmental regulations are driving increased adoption of WHR systems. As regulations on greenhouse gas emissions become more stringent, WHR will become an increasingly essential technology for compliance. This will create new opportunities for manufacturers of WHR equipment and drive further innovation in this field. I believe that WHR is not just a trend; it's a fundamental shift in how we think about energy utilization in the maritime industry.

6 Chapter 6: Control Systems Integration

Marine vessels are complex systems that require precise control to operate efficiently and safely. Control systems integration is the art and science of orchestrating these disparate systems into a harmonious whole. The propulsion system, the heart of the vessel, must be carefully managed to deliver the desired speed and thrust, while also minimizing fuel consumption and emissions. This chapter will shed light on how control systems achieve this delicate balance.

Automatic speed control (ASC) is a cornerstone of efficient marine propulsion. Imagine manually adjusting the engine throttle every few seconds to maintain a constant speed, battling wind, waves, and currents. ASC systems automate this process, freeing the crew to focus on other critical tasks. These systems typically rely on a closed-loop feedback mechanism. A speed sensor measures the vessel's actual speed, and this value is compared to the desired speed set by the operator. Any difference between the two, known as the error signal, is fed into a controller. The controller then adjusts the engine throttle or propeller pitch to reduce the error, bringing the vessel closer to the desired speed. Sophisticated ASC systems consider not only the speed error but also its rate of change and accumulated value over time. This allows the controller to anticipate future speed changes and make smoother, more responsive adjustments. For instance, if the vessel encounters a headwind, the ASC system will proactively increase engine power to prevent a significant speed drop. Furthermore, modern ASC systems often incorporate adaptive control algorithms that learn the vessel's dynamic characteristics over time. This allows them to compensate for changes in hull fouling, load conditions, and environmental factors, ensuring optimal performance under a wide range of operating conditions. The benefits of automatic speed control extend beyond simple convenience. By maintaining a more consistent speed, ASC systems reduce fuel consumption and emissions, contributing to a more sustainable maritime industry. They also improve safety by reducing the workload on the crew and allowing them to focus on navigation and collision avoidance.

Dynamic positioning systems (DPS) represent a significant leap forward in marine vessel control. Consider the challenge of maintaining a precise location in the open ocean, subject to the forces of wind, waves, and currents. Traditional mooring systems are often impractical or impossible in deep water or harsh environments. DPS provides a solution by using a combination of sensors, thrusters, and sophisticated control algorithms to automatically maintain the vessel's position and heading. The heart of a DPS is its control system, which continuously monitors the vessel's position and heading using a variety of sensors, including

GPS, gyrocompasses, and motion reference units. This information is then fed into a mathematical model that estimates the forces acting on the vessel. Based on this estimate, the control system calculates the thrust required from each thruster to counteract these forces and maintain the desired position and heading. The thrusters, which can be either fixed-pitch or controllable-pitch propellers, are then activated to generate the required thrust. DPS can operate in a variety of modes, depending on the application. In position-keeping mode, the system maintains the vessel at a fixed location. In track-following mode, the system guides the vessel along a predetermined path. In heading-control mode, the system maintains the vessel's heading, regardless of external forces. DPS are essential for a wide range of offshore operations, including oil and gas exploration, subsea construction, and scientific research. They allow vessels to operate safely and efficiently in challenging environments, enabling tasks that would be impossible with traditional mooring systems. The accuracy and reliability of DPS are critical for these operations, as even small errors in position or heading can have significant consequences. Therefore, DPS are designed with multiple layers of redundancy to ensure that the system remains operational even in the event of a component failure.

Integrated bridge systems (IBS) represent the pinnacle of control systems integration on modern marine vessels. Think of the bridge of a traditional ship, with its array of separate instruments and controls for navigation, communication, and propulsion. An IBS consolidates these functions into a single, integrated platform, providing the bridge team with a comprehensive overview of the vessel's status and its surrounding environment. At the heart of an IBS is a central computer system that collects data from various sensors and systems throughout the vessel. This data is then processed and displayed on a series of high-resolution screens, providing the bridge team with a unified view of the vessel's position, speed, heading, engine status, and other critical parameters. The IBS also integrates various control functions, allowing the bridge team to control the vessel's propulsion system, steering gear, and other equipment from a single location. This simplifies the operation of the vessel and reduces the workload on the crew. Moreover, IBS often incorporate advanced decision support tools, such as electronic chart display and information systems (ECDIS), radar systems, and automatic identification systems (AIS). These tools provide the bridge team with real-time information about the vessel's surroundings, helping them to make informed decisions and avoid potential hazards. IBS are designed to improve safety, efficiency, and situational awareness on marine vessels. By consolidating control functions and providing the bridge team with a comprehensive overview of the vessel's status, IBS reduce the risk of human error and improve the overall efficiency of operations. They also enhance situational awareness by providing the bridge team with real-time information about the vessel's surroundings, allowing them to anticipate potential hazards and take proactive measures to avoid them. Ultimately, the integration of control systems into marine vessels is a continuous process of innovation and refinement. As technology advances, we can expect to see even more sophisticated and integrated systems that further enhance the safety, efficiency, and sustainability of maritime operations.

6.1 Automatic Speed Control

Maintaining a consistent speed is paramount for several reasons. Firstly, it directly impacts fuel efficiency. Sudden acceleration and deceleration require significantly more energy than maintaining a steady pace. Secondly, speed consistency contributes to passenger comfort and cargo stability, especially in challenging sea conditions. Thirdly, predictable speeds are crucial for accurate navigation and adherence to schedules, vital aspects of commercial shipping and passenger transport. Finally, consistent speed management reduces wear and tear on propulsion machinery, prolonging its lifespan and minimizing maintenance costs. Automatic speed control systems are designed to address these needs by precisely regulating the engine's output to maintain a desired speed, irrespective of external disturbances like wind, waves, and changing hull resistance.

At its core, an automatic speed control (ASC) system for marine vessels operates on the principle of closed-loop feedback control. A speed sensor, typically a shaft speed encoder or GPS-based system, continuously monitors the vessel's actual speed and transmits this information to a control unit. The control unit, often a programmable logic controller (PLC) or a dedicated microprocessor-based system, compares the actual speed to the desired speed, which is pre-set by the operator or determined by a higher-level navigation system. The difference between the desired speed and the actual speed generates an error signal. This error signal is then used by a control algorithm to adjust the engine's fuel injection rate, throttle position, or propeller pitch (in the case of controllable pitch propellers), thereby altering the engine's output torque and ultimately influencing the vessel's speed. The control loop continuously iterates, making real-time adjustments to maintain the desired speed as accurately as possible.

The control algorithm is the heart of any automatic speed control system. Proportional-Integral-Derivative (PID) controllers are widely used in marine applications due to their robustness and ease of tuning. The proportional component of the PID controller provides a control action that is proportional to the speed error. A larger error results in a larger corrective action. The integral component eliminates steady-state errors by accumulating the error over time and applying a corrective action to drive the error to zero. The derivative component anticipates future errors by responding to the rate of change of the error. This helps to dampen oscillations and improve the system's response time. Tuning the PID controller involves carefully selecting the gains for each component (proportional gain, integral gain, and derivative gain) to achieve the desired performance characteristics, such as fast response time, minimal overshoot, and good stability. More advanced control algorithms, such as adaptive control and model predictive control, can also be used to improve performance, particularly in situations where the vessel's dynamics are highly nonlinear or time-varying. These algorithms can adapt to changing operating conditions and optimize the control action based on a model of the vessel's behavior.

Implementing an automatic speed control system requires careful consideration of several factors. The selection of the speed sensor is crucial for accurate speed measurement. Shaft speed encoders provide high-resolution speed measurements but are susceptible to noise and vibration. GPS-based systems offer accurate speed measurements but may be affected by signal loss in certain areas. The choice of the control unit depends on the complexity of the control algorithm and the required processing power. PLCs are commonly used for simpler control systems, while microprocessors or dedicated controllers are preferred for more complex systems. The actuator, which is responsible for adjusting the engine's output, must be carefully selected to ensure that it can respond quickly and accurately to the control signals. In addition, the system must be designed to be robust and reliable, capable of withstanding the harsh marine environment. This includes protecting the system from corrosion, vibration, and extreme temperatures. Redundancy and fault tolerance are also important considerations for critical applications.

Practical operation of an ASC system involves several key steps. First, the system must be properly calibrated and tuned to ensure that it is functioning correctly. This typically involves running a series of tests and adjusting the controller parameters until the desired performance is achieved. Once the system is calibrated, the operator can set the desired speed using a control panel or a navigation system. The ASC system will then automatically adjust the engine's output to maintain the desired speed, even in the face of disturbances. During operation, it is important to monitor the system's performance and make adjustments as needed. This may involve adjusting the controller parameters to compensate for changes in the vessel's loading condition or environmental conditions. It is also important to perform regular maintenance on the system to ensure that it continues to function reliably. This includes inspecting the sensors, actuators, and control unit for signs of wear and tear, and replacing any components that are damaged or worn out.

Several advancements are shaping the future of automatic speed control in marine applications. One significant trend is the increasing use of data analytics and machine learning to optimize control performance. By analyzing historical data on vessel speed, engine performance, and environmental conditions, it is possible to develop models that can predict the vessel's behavior and optimize the control action in real time. Another trend is the integration of ASC systems with other vessel systems, such as navigation systems, energy management systems, and propulsion control systems. This integration allows for more coordinated and efficient operation of the vessel. For example, the ASC system can be integrated with a navigation system to automatically adjust the vessel's speed based on the planned route and weather conditions. Furthermore, we can expect a gradual shift towards more electric propulsion systems, where speed control is intimately linked to managing electrical power flow within the vessel, offering further opportunities for efficiency gains. Finally, the integration of remote monitoring and control capabilities allows for shore-based operators to monitor and control the vessel's speed remotely, further enhancing efficiency and safety.

6.2 Dynamic Positioning Systems

Dynamic Positioning (DP) systems represent a cornerstone in modern marine operations, particularly for tasks demanding utmost precision and stability. Understanding their functionality is crucial for anyone involved in the design, operation, or maintenance of advanced marine vessels, which makes it highly important to include this chapter to the book. At their core, DP systems are sophisticated control networks that automatically maintain a vessel's position and heading using its own propellers and thrusters. This capability is invaluable in scenarios where traditional anchoring is impractical or impossible, such as deep-sea drilling, offshore construction, and cable laying. The level of control and accuracy provided by DP systems are essential to ensure safe and efficient operations under challenging environmental conditions.

The fundamental principle behind DP relies on a closed-loop feedback system. This involves several key components working in harmony. Firstly, sensors constantly monitor the vessel's position and heading. These sensors can include GPS, laser rangefinders, gyrocompasses, motion reference units (MRUs) and wind sensors, providing a comprehensive overview of the vessel's spatial orientation and environmental influences. The data from these sensors is fed into a sophisticated control computer. This computer acts as the brain of the DP system, processing the sensor data and comparing it to the desired position and heading set by the operator. The control computer then calculates the necessary thrust adjustments needed to counteract any deviations from the desired position. Finally, based on the calculations, the control computer sends commands to the vessel's thrusters and propellers, adjusting their speed and direction to maintain the desired position and heading. This continuous feedback loop ensures that the vessel remains stable despite external forces like wind, waves, and currents.

The reliability and redundancy of DP systems are paramount, particularly in safety-critical applications. DP systems are classified into different equipment classes (DP0, DP1, DP2, DP3) based on their redundancy and fault tolerance capabilities, with higher classes indicating greater levels of redundancy. For instance, a DP2 system is designed to withstand a single fault without losing position, while a DP3 system can withstand even more severe failures, such as a fire or flood in a single compartment. Redundancy is achieved through multiple independent systems for power, control, and thrust. This means having multiple generators, switchboards, controllers, and thrusters, each capable of taking over in case of a failure in another system. For example, if one generator fails, another generator automatically starts up to provide the necessary power. Similarly, if one thruster malfunctions, other thrusters can compensate to maintain the vessel's position. The design and implementation of redundancy are governed by strict industry standards, such as those set by IMO (International Maritime Organization) and classification societies like DNV GL and ABS. These standards ensure that DP systems meet specific safety and performance requirements based on the intended operational environment and risk assessment.

Operating a DP system effectively requires a deep understanding of its capabilities and limitations, as well as a comprehensive awareness of the environmental conditions. DP operators must be properly trained and certified, possessing the skills to interpret sensor data, assess risk, and respond to potential failures. Before commencing DP operations, a thorough risk assessment should be conducted, considering factors such as weather conditions, sea state, proximity to other structures, and potential consequences of a position loss. The DP operator must also be proficient in setting up the DP system, configuring its parameters, and monitoring its performance. During operations, continuous monitoring of sensor data, thrust allocation, and environmental conditions is essential. The operator must be prepared to take manual control of the vessel if necessary and to implement contingency plans in the event of a system failure. Furthermore, communication with other parties involved in the operation, such as the bridge team, engineers, and clients, is crucial for maintaining situational awareness and ensuring a coordinated response to any unforeseen events.

Maintaining the optimal performance of DP systems involves regular inspections, testing, and maintenance. Key components, such as sensors, controllers, thrusters, and power systems, should be inspected and maintained according to the manufacturer's recommendations and industry best practices. Regular testing of redundancy features is essential to verify their functionality and ensure that they will perform as expected in the event of a failure. This includes testing the automatic start-up of backup generators, the switchover of control systems, and the ability of redundant thrusters to compensate for a failed thruster. Furthermore, software updates and upgrades should be installed promptly to address any known issues and to take advantage of new features and improvements. Proper documentation of all maintenance activities is crucial for tracking the history of the system and identifying any potential problems. A well-documented maintenance program helps to ensure the long-term reliability and availability of the DP system, reducing the risk of failures and minimizing downtime.

The integration of DP systems with other vessel systems is becoming increasingly common, leading to enhanced functionality and efficiency. Integrating DP with the vessel's power management system can optimize energy consumption by automatically adjusting the power output of generators based on the actual thrust demand. This reduces fuel consumption and emissions, contributing to a more sustainable operation. Similarly, integrating DP with the vessel's navigation system allows for more precise route planning and execution, particularly in congested waters or areas with restricted access. Furthermore, integrating DP with environmental monitoring systems can provide real-time information on wind, waves, and currents, allowing the DP system to adapt its control strategy to minimize the impact of these external forces. This integration of systems not only improves the performance of the DP system but also enhances the overall efficiency and safety of the vessel operation.

The future of DP systems is likely to be shaped by advancements in technology and the increasing demands of the offshore industry. One trend is the development of more intelligent DP systems that can learn from past experiences and adapt their control strategies to optimize performance in different environmental conditions. Another trend is the integration of DP systems with remote monitoring and control capabilities, allowing for remote operation of vessels and reduced crew sizes. In addition, there is growing interest in using renewable energy sources, such as wind and solar power, to power DP systems, reducing their reliance on fossil fuels. Finally, the development of new thruster technologies, such as electric thrusters and rim-driven thrusters, promises to improve the efficiency and reliability of DP systems. These advancements will enable DP systems to operate in more challenging environments, perform more complex tasks, and contribute to a more sustainable marine industry. I know this is a crucial system to understand in modern maritime.

6.3 Integrated Bridge Systems

Integrated Bridge Systems (IBS) represent a significant advancement in marine navigation and operational efficiency. They move beyond individual, stand-alone pieces of equipment to create a cohesive, integrated platform that provides a holistic view of the vessel's status and its surrounding environment. This integration simplifies tasks for bridge personnel, reduces workload, and ultimately enhances safety and decision-making. We can see integrated bridge systems in every ship today.
The core concept of an IBS revolves around centralizing control and monitoring functions. Rather than having separate displays and control panels for radar, ECDIS (Electronic Chart Display and Information System), autopilot, communication systems, and en-

gine monitoring, an IBS consolidates these functions into a unified workstation or a network of interconnected workstations. This allows bridge officers to access and interact with all critical data from a single point, streamlining operations and improving situational awareness. Imagine a scenario where a sudden course alteration is needed to avoid an obstacle detected by radar; with an IBS, the officer can immediately adjust the autopilot settings, update the ECDIS route, and communicate the change to relevant parties, all from a single interface. The use of these systems reduces the chance of errors in communication.

Several key components typically comprise an IBS. The ECDIS forms a crucial part, displaying electronic navigational charts and integrating real-time data such as GPS position, radar overlay, and Automatic Identification System (AIS) targets. The radar system, often with advanced features like automatic target tracking (ATA) and collision avoidance systems (ARPA), provides crucial information about surrounding vessels and obstacles, especially in low-visibility conditions. An autopilot system allows for precise course keeping and automated navigation, reducing the workload on the helmsman and improving fuel efficiency. The communication systems, including VHF radios, satellite communication devices, and internal communication networks, ensure effective communication with shore-based authorities, other vessels, and onboard crew members. A central alarm and monitoring system (CAMS) continuously monitors various parameters, such as engine performance, tank levels, and fire detection systems, providing early warnings of potential problems. These components are not just separately existing; rather, they exist for a purpose.

Implementing an IBS offers numerous advantages. Enhanced situational awareness is a primary benefit, as the integrated display of data from various sensors provides a more complete and accurate picture of the vessel's surroundings. This improved awareness allows bridge officers to make more informed decisions, especially in complex or hazardous situations. Reduced workload is another significant advantage, as the centralized control and monitoring functions streamline operations and reduce the need for manual data entry and interpretation. This frees up bridge officers to focus on other critical tasks, such as watchkeeping and collision avoidance. Improved safety is a direct consequence of the enhanced situational awareness and reduced workload, as the IBS helps to prevent accidents and improve the overall safety of navigation. Enhanced efficiency is also achieved through the optimized routing, speed control, and fuel management capabilities offered by the IBS, leading to reduced fuel consumption and lower operating costs. Integrated bridge systems are not a luxury, they are a requirement.

However, the implementation of an IBS also presents several challenges. The initial cost of installing an IBS can be significant, requiring a substantial investment in hardware, software, and training. The complexity of the system can also be a challenge, requiring thorough training and familiarization for bridge officers to operate it effectively. Integration issues between different components can arise, especially when integrating equipment from different manufacturers. Ensuring compatibility and seamless data exchange between different systems is crucial for the proper functioning of the IBS. Reliance on automation can also be a potential drawback, as bridge officers must maintain their manual skills and be prepared to take over control in the event of system failures. Cybersecurity threats are another growing concern, as the interconnected nature of the IBS makes it vulnerable to cyberattacks that could compromise the vessel's navigation and safety. I have seen a lot of captain lose their focus on this fact.

Successfully integrating and operating an IBS requires careful planning and execution. A comprehensive risk assessment should be conducted to identify potential hazards and vulnerabilities associated with the IBS. A detailed training program should be developed to ensure that bridge officers are proficient in the operation and maintenance of the system. Regular maintenance and testing should be performed to ensure that the IBS is functioning properly and that all components are calibrated and up to date. Redundancy should be built into the system, with backup systems and manual overrides available in case of failures. Cybersecurity measures should be implemented to protect the IBS from cyberattacks, including firewalls, intrusion detection systems, and regular security audits. As someone who has been in the industry, if you don't make this considerations you're doomed.

Looking ahead, future developments in IBS technology are likely to focus on further integration, automation, and data analysis. The integration of augmented reality (AR) and virtual reality (VR) technologies could provide bridge officers with enhanced situational awareness and decision support. The use of artificial intelligence (AI) and machine learning (ML) algorithms could enable the IBS to automatically optimize routing, speed control, and fuel consumption based on real-time data and predictive models. The development of more secure and resilient communication networks will be crucial for protecting the IBS from cyberattacks. As technology advances, IBS will continue to play an increasingly important role in ensuring the safety, efficiency, and sustainability of marine operations. The possibilities are endless, and it is up to us to seize them.

In conclusion, Integrated Bridge Systems are a vital component in modern marine craft, offering numerous benefits in terms of safety, efficiency, and situational awareness. However, successful implementation requires careful planning, thorough training, and ongoing maintenance. As technology continues to evolve, IBS will become even more sophisticated and integrated, further enhancing the capabilities of marine vessels and contributing to a safer and more sustainable maritime industry. This is not merely a technological advancement; it is a paradigm shift in how we operate at sea. Without making an effort you will not achieve your goals.

7 Chapter 7: Future Trends in Propulsion

The seas are changing, and so too are the ways we navigate them. For centuries, marine propulsion has relied on variations of established technologies. But the winds of change are blowing, propelled by escalating environmental concerns, stringent regulations, and the relentless pursuit of efficiency. These factors are not just nudging the industry; they are fundamentally reshaping its future. Now, it's time to set our sights on the horizon and examine the key trends that will define the next era of marine propulsion.

Electric and hybrid propulsion systems are gaining significant momentum, driven by the dual imperatives of reducing emissions and improving fuel efficiency. These systems represent a departure from traditional reliance on internal combustion engines, instead integrating electric motors, batteries, and sometimes, generators powered by cleaner fuels. Consider the operational profile of various vessel types. Ferries operating on fixed routes with frequent port calls are ideal candidates for full electrification, leveraging shore power for charging and eliminating emissions within urban areas. On the other hand, larger vessels engaged in long-haul voyages may benefit more from hybrid systems, where electric propulsion assists during maneuvering and low-speed operations, with traditional engines providing power for high-speed transits. The specific architecture of these systems varies considerably, with series hybrids, parallel hybrids, and fully electric configurations each offering distinct advantages in terms of efficiency, power delivery, and operational flexibility. Battery technology remains a critical factor, with ongoing advancements in energy density, charging rates, and safety improving the viability of electric propulsion for a wider range of vessel applications. Regulatory support, through incentives and emission control mandates, is further accelerating the adoption of these systems.

Autonomous vessel technology is rapidly evolving, promising to revolutionize maritime operations through enhanced safety, efficiency, and cost-effectiveness. While fully autonomous vessels remain some way off, the integration of automation technologies into existing propulsion systems is already transforming vessel management. One example is the development of advanced sensor systems that provide real-time data on vessel performance, environmental conditions, and potential hazards. These data are then fed into sophisticated control algorithms that optimize engine settings, route planning, and speed adjustments to minimize fuel consumption and emissions. Autonomous navigation systems can also improve safety by reducing the risk of human error, particularly in congested waterways or adverse weather conditions. Remote monitoring and control centers allow shore-based operators to oversee vessel operations, providing support and intervention when necessary. The legal and regulatory frameworks surrounding autonomous vessels are still under development, but pilot projects and trials are paving the way for wider adoption of this technology. The integration of propulsion systems with autonomous control algorithms is a crucial area of research, focusing on developing robust and reliable systems that can operate safely and efficiently in a variety of conditions.

The search for sustainable fuel alternatives is becoming increasingly urgent as the maritime industry strives to decarbonize its operations. Traditional fossil fuels are major contributors to greenhouse gas emissions and air pollution, prompting a global effort to identify and adopt cleaner energy sources. Liquefied natural gas (LNG) is already widely used as a transition fuel, offering lower emissions compared to heavy fuel oil. However, longer-term solutions are needed to achieve truly sustainable operations. Biofuels, derived from renewable sources such as algae, vegetable oils, and agricultural waste, offer the potential for significant emissions reductions. However, challenges remain in terms of scalability, cost, and environmental impact of biofuel production. Hydrogen is another promising alternative fuel, with the potential to produce zero emissions when used in fuel cells or internal combustion engines. However, the production, storage, and distribution of hydrogen present significant technical and logistical challenges. Ammonia, a carbon-free compound that can be produced from renewable sources, is also gaining attention as a potential marine fuel. Each fuel faces unique technical hurdles related to engine design, fuel storage, and safety protocols. Developing the infrastructure needed to support these alternative fuels is a critical step towards their widespread adoption.

Advanced materials are playing an increasingly important role in improving the performance and durability of marine propulsion systems. Lighter and stronger materials can reduce the weight of engine components, leading to improved fuel efficiency. High-performance alloys can withstand higher temperatures and pressures, enabling the development of more powerful and efficient engines. Composite materials, such as carbon fiber reinforced polymers, are being used to manufacture propellers and other components, reducing weight and improving hydrodynamic performance. Nanomaterials are also being explored for use in coatings and lubricants, reducing friction and wear. Additive manufacturing, also known as 3D printing, is revolutionizing the way propulsion components are designed and manufactured, enabling the creation of complex geometries and customized parts. These new materials and manufacturing techniques are not only improving the performance of existing propulsion systems but also enabling the development of entirely new propulsion concepts.

The integration of sophisticated learning models offers transformative possibilities for propulsion control, promising optimized energy efficiency and enhanced safety protocols. Picture intelligent systems analyzing real-time data from an array of sensors - engine performance, hull condition, weather patterns, and navigational information - to dynamically adjust propulsion parameters. This might mean fine-tuning engine settings to minimize fuel consumption based on current sea state and vessel load, or proactively adjusting course and speed to avoid adverse weather conditions. Moreover, these systems can learn from historical data, identifying patterns and anomalies that could indicate potential maintenance needs or safety hazards. Predictive maintenance algorithms can anticipate component failures, enabling proactive repairs and minimizing downtime. These systems can also enhance safety by detecting and responding to potential collisions or groundings, providing alerts and suggesting corrective actions. The deployment of such sophisticated control systems requires robust data infrastructure, secure communication networks, and skilled personnel to manage and interpret the data. As these technologies mature, they will play a crucial role in optimizing the performance and safety of marine propulsion systems, paving the way for a more efficient and sustainable maritime industry.

Ultimately, the future of marine propulsion is not defined by any single technology but by the convergence of these trends. Electric and hybrid systems, autonomous controls, sustainable fuels, and advanced materials are all interconnected, each influencing and accelerating the development of the others. Navigating this complex landscape requires a holistic approach, considering the environmental, economic, and social implications of each technology. Embracing innovation and fostering collaboration among researchers, industry stakeholders, and policymakers will be essential to unlocking the full potential of these future trends and creating a cleaner, more efficient, and more sustainable maritime industry. The journey is underway, and the destination is a future where our vessels glide through the oceans with minimal impact, powered by the ingenuity of human innovation.

7.1 Electric and Hybrid Propulsion

The whispers of change in marine propulsion are growing louder, and at the forefront of this revolution are electric and hybrid systems. For too long, the maritime industry has relied heavily on fossil fuels, contributing significantly to global emissions. Now, with increasing environmental awareness and tightening regulations, the adoption of cleaner, more sustainable propulsion methods is not just desirable but essential. Electric and hybrid propulsion systems offer a promising path towards reducing the environmental footprint of marine vessels while simultaneously improving operational efficiency and reducing fuel costs in the long run, so don't think of them as a far-off future; they are a present necessity.

Let's start with the fundamentals. An electric propulsion system relies entirely on electricity to drive the vessel's propeller or other propulsion devices. This electricity can be generated from various sources, including batteries, fuel cells, or shore power. Hybrid propulsion, on the other hand, combines an electric system with a traditional internal combustion engine (ICE). The ICE typically acts as a generator, producing electricity to power the electric motor and charge batteries. In some hybrid configurations, the ICE can directly drive the propeller, providing additional power during high-speed operation. One of the main differentiators lies in operational flexibility. Electric systems are remarkably silent and produce zero emissions at the point of operation, which is ideal for environmentally sensitive areas like ports and coastal zones. However, their range is typically limited by battery capacity. Hybrid systems offer greater range and operational flexibility, allowing vessels to switch between electric and ICE modes depending on the situation. During low-speed maneuvering or when entering port, the vessel can operate in electric mode, minimizing emissions and noise. At higher speeds or during long voyages, the ICE can be engaged to extend the vessel's range.

Now, let's discuss the various configurations you might encounter. Series hybrid systems feature an engine driving a generator, which in turn powers an electric motor connected to the propeller. There's no direct mechanical link between the engine and the propeller. This configuration offers excellent control and efficiency, as the engine can always operate at its optimal speed, regardless of the vessel's speed. Parallel hybrid systems, on the other hand, allow both the engine and the electric motor to directly drive the propeller. This provides greater flexibility and allows the vessel to utilize the combined power of both systems when needed. A third option, the series-parallel hybrid system, combines the features of both series and parallel configurations. This configuration offers the greatest flexibility and efficiency, but it is also the most complex. Understanding these configurations is crucial when evaluating the suitability of a hybrid system for a specific vessel.

Looking at the core components, you'll find the battery system forms the heart of any electric or hybrid propulsion system. Batteries store the electrical energy that powers the electric motor. Lithium-ion batteries are the most commonly used type due to their high energy density and relatively long lifespan. However, other battery technologies, such as solid-state batteries, are being developed and could offer even better performance in the future. The electric motor is responsible for converting electrical energy into mechanical energy to drive the propeller. Electric motors are highly efficient and offer excellent torque characteristics, making them well-suited for marine propulsion. Synchronous motors and induction motors are two common types used in marine applications. The generator, found in hybrid systems, converts mechanical energy from the engine into electrical energy. It is typically a synchronous generator, known for its reliability and efficiency. Power electronics, including inverters and converters, are essential for controlling the flow of electrical energy within the propulsion system. Inverters convert DC electricity from the batteries or generator into AC electricity for the electric motor, while converters regulate the voltage and current levels.

Delving into efficiency considerations, it's crucial to understand how these systems contribute to fuel savings and reduced emissions. Electric and hybrid propulsion systems can significantly improve overall vessel efficiency compared to traditional ICE-based systems. Electric motors are inherently more efficient than ICEs, and hybrid systems allow the engine to operate at its optimal efficiency point, further reducing fuel consumption. Moreover, the ability to operate in electric mode in environmentally sensitive areas eliminates emissions altogether. But remember, the overall efficiency also depends on factors such as the vessel's operating profile, the type of battery and motor used, and the effectiveness of the energy management system.

Moving onto practical applications, you'll find that electric and hybrid propulsion systems are finding increasing adoption across a wide range of vessel types. Ferries and passenger vessels are particularly well-suited for electric and hybrid propulsion, as they typically operate on fixed routes with frequent stops, allowing for battery recharging. Tugboats and workboats benefit from the high torque and maneuverability offered by electric motors. Offshore support vessels are also increasingly adopting hybrid propulsion to reduce emissions and improve fuel efficiency during operations. Even larger vessels, such as container ships and cruise ships, are beginning to explore hybrid propulsion as a means of meeting increasingly stringent environmental regulations.

Let's not forget the challenges and considerations that come with these systems. One of the main challenges is the initial cost of electric and hybrid propulsion systems, which can be higher than traditional ICE-based systems. However, this cost can be offset by reduced fuel consumption, lower maintenance costs, and potential government incentives. Battery technology is also an area of ongoing development. Improving energy density, lifespan, and safety are crucial for expanding the adoption of electric propulsion. Furthermore, the availability of charging infrastructure, especially in ports, is essential for supporting electric vessels.

Finally, let's look ahead. The future of electric and hybrid propulsion in the marine industry is bright. As battery technology continues to improve and regulations become stricter, the adoption of these systems is expected to accelerate. We can anticipate seeing larger vessels with fully electric propulsion, powered by advanced battery systems or fuel cells. The integration of renewable energy sources, such as solar and wind, will further enhance the sustainability of marine propulsion. Furthermore, advanced control systems and energy management algorithms will play a key role in optimizing the efficiency and performance of electric and hybrid vessels. By embracing these technologies, the maritime industry can move towards a cleaner, more sustainable future.

7.2 Autonomous Vessel Technology

The allure of the open sea has always been interwoven with humanity's ingenuity, a relentless drive to push the boundaries of what's possible. We've witnessed incredible transformations in marine propulsion, from the age of sail to the era of powerful diesel engines. Now, we stand at the cusp of another revolution: autonomous vessel technology, a field promising to redefine maritime operations as we know them. For a beginner, the sheer complexity of this field can be daunting. However, the core concepts are surprisingly accessible, and understanding them is crucial for anyone interested in the future of maritime transport. Autonomous vessel technology isn't simply about building ships that sail themselves; it's a complex interplay of numerous interconnected systems working in perfect harmony. I always tell aspiring engineers to think of it as a symphony orchestra where each instrument represents a critical technology: advanced sensors, sophisticated control algorithms, robust communication networks, and reliable propulsion systems. The conductor? A powerful onboard computer capable of processing vast amounts of data and making real-time decisions. At the heart of this system lies perception. An autonomous vessel needs to "see" its surroundings, and that's where sensors come into play. Radar provides long-range detection of other vessels and obstacles, while lidar (Light Detection and Ranging) creates detailed 3D maps of the environment. Cameras, both visible light and infrared, offer visual information, and sonar helps navigate underwater. The data from these sensors is then fed into a central processing unit, the "brain" of the vessel. This is where the magic happens. Algorithms analyze the sensor data, identifying potential hazards, calculating optimal routes, and making decisions on speed and course corrections. These algorithms must be incredibly robust, capable of handling unexpected events and adapting to changing conditions.

Navigation is, understandably, key. A sophisticated GPS system provides precise location information, but GPS alone isn't enough. To ensure accuracy and reliability, autonomous vessels often use inertial navigation systems (INS), which rely on accelerometers and gyroscopes to track the vessel's movement relative to its starting point. Combining GPS with INS allows the vessel to maintain its position even when GPS signals are temporarily unavailable. Communication is another vital aspect. Autonomous vessels need to communicate with shore-based control centers, other vessels, and potentially even satellites. This requires a reliable communication network, often involving a combination of satellite links, radio communication, and cellular networks. Redundancy is crucial here; the vessel needs to maintain communication even if one or more communication channels fail.

Now, let's think about power. The reliability of the propulsion system is paramount. Imagine an autonomous vessel stranded in the middle of the ocean due to engine failure. Redundancy is again essential. Many autonomous vessels are equipped with multiple propulsion systems, allowing them to continue operating even if one system malfunctions. Regular maintenance and rigorous testing are also critical to ensuring the long-term reliability of the propulsion system. I once worked on a project where we implemented a predictive maintenance system that used sensor data to detect potential problems before they occurred, reducing the risk of unexpected breakdowns.

Beyond the technical aspects, regulations and ethical considerations play a significant role in the development and deployment of autonomous vessels. The legal framework governing autonomous shipping is still evolving. Questions of liability in the event of an accident, insurance requirements, and cybersecurity protocols are being actively debated and developed. What happens if an autonomous vessel faces a situation where any decision it makes will result in harm? Who is responsible? These are difficult questions with no easy answers. From my experience, many companies are investing heavily in cybersecurity to protect their autonomous vessels from cyberattacks. After all, a hacked autonomous vessel could pose a significant threat to maritime safety.

Finally, think about the human element. The introduction of autonomous vessels will undoubtedly have a significant impact on the maritime workforce. While some jobs may be lost, new opportunities will also be created in areas such as software development, sensor maintenance, and remote monitoring. Retraining and upskilling programs will be essential to ensure that maritime workers have the skills needed to thrive in this changing landscape. We should embrace these changes and look for ways to integrate autonomous technology into existing maritime operations, to improve safety, efficiency, and sustainability.

7.3 Sustainable Fuel Alternatives

The quest for sustainable fuel alternatives in the maritime industry is no longer a far-off aspiration; it's a rapidly evolving necessity, driven by tightening environmental regulations, increasing fuel costs, and a growing societal awareness of the impact of shipping on the planet. Finding fuels that don't contribute to greenhouse gas emissions, reduce air pollution, and can be readily integrated into existing infrastructure presents a significant challenge, but also a tremendous opportunity for innovation. This chapter will explore some of the most promising alternatives, examining their current state of development, potential benefits, and the hurdles that need to be overcome for widespread adoption.

Liquefied Natural Gas (LNG) represents a significant step towards cleaner maritime operations. Primarily composed of methane, LNG combustion produces significantly lower levels of sulfur oxides (SOx), particulate matter (PM), and nitrogen oxides (NOx) compared to traditional heavy fuel oil (HFO). Moreover, it offers a reduction in greenhouse gas emissions, although the precise amount depends on the engine technology used and the potential for methane slip - the release of uncombusted methane into the atmosphere, a potent greenhouse gas. While LNG infrastructure is developing globally, its availability remains limited in certain regions, requiring substantial investment in new bunkering facilities and port infrastructure. Retrofitting existing vessels to use LNG can be a complex and costly undertaking, often involving significant modifications to the engine, fuel storage, and safety systems. However, for new builds, LNG offers a viable and increasingly attractive option, particularly for vessels operating in emission control areas (ECAs) where stringent limits on SOx and NOx emissions are enforced. The long-term sustainability of LNG also depends on the source of the natural gas. While conventional natural gas is a fossil fuel, biogas and synthetic natural gas (produced from renewable sources) offer pathways to even greater reductions in greenhouse gas emissions.

Methanol and ethanol, both alcohols, present compelling alternatives with different production pathways and characteristics. Methanol can be produced from natural gas, coal, or renewable sources such as biomass and captured carbon dioxide. Ethanol is primarily produced from biomass, such as corn, sugarcane, or cellulosic feedstocks. Both fuels offer the potential for significant reductions in SOx, PM, and NOx emissions, and can achieve substantial greenhouse gas reductions depending on their production method. Methanol, in particular, has gained traction as a marine fuel due to its relatively low cost, ease of handling, and compatibility with existing bunkering infrastructure with some modifications. Several vessels are already operating on methanol, demonstrating its technical feasibility. However, both methanol and ethanol have lower energy densities than HFO, meaning that vessels need to carry larger volumes of fuel to achieve the same range. They also require modifications to engine design and fuel storage systems to ensure safe and efficient operation. The sustainability of methanol and ethanol depends heavily on the feedstock and production process. Production from renewable sources, such as biomass or captured carbon dioxide, offers the greatest potential for greenhouse gas reductions, but scaling up these production methods to meet the demands of the global shipping industry remains a challenge.

Hydrogen, the simplest and most abundant element in the universe, is increasingly viewed as a long-term solution for zero-emission shipping. When used in a fuel cell, hydrogen reacts with oxygen to produce electricity, with water as the only byproduct. Hydrogen can also be combusted in an internal combustion engine, although this method typically produces lower efficiency and some NOx emissions. The primary challenge with hydrogen is its production, storage, and distribution. Currently, most hydrogen is produced from natural gas through a process called steam methane reforming, which generates significant carbon dioxide emissions. However, hydrogen can also be produced through electrolysis, using electricity to split water into hydrogen and oxygen. If the electricity comes from renewable sources, such as solar or wind power, the resulting hydrogen is considered "green hydrogen" and is virtually carbon-free. Storing hydrogen onboard vessels is another major hurdle. Due to its low volumetric energy density, hydrogen must be either compressed to very high pressures, liquefied at extremely low temperatures (-253°C), or chemically bound to a carrier molecule such as ammonia. Each storage method has its own advantages and disadvantages in terms of energy density, cost, and safety. Despite these challenges, significant research and development efforts are underway to advance hydrogen technology for marine applications, with pilot projects demonstrating the feasibility of hydrogen-powered vessels.

Ammonia (NH3) emerges as another promising hydrogen carrier and potential fuel in its own right. Ammonia is a compound of nitrogen and hydrogen that can be produced from renewable sources, using electrolysis to produce hydrogen and then combining it with nitrogen from the air. While ammonia combustion does produce NOx emissions, these can be mitigated through advanced engine technologies. A key advantage of ammonia is its higher volumetric energy density compared to hydrogen, making it easier to store and transport. Ammonia also has an established global distribution network, as it is widely used as a fertilizer. However, ammonia is toxic and corrosive, requiring stringent safety measures for handling and storage. Research is ongoing to develop ammonia-fueled engines and fuel cells that can operate safely and efficiently in marine environments.

Biofuels, derived from renewable biomass sources such as algae, vegetable oils, and agricultural residues, offer a drop-in replacement for conventional fossil fuels in many applications. Biodiesel can be used in existing diesel engines with minimal modifications, while renewable diesel offers even better performance and compatibility. Biofuels can significantly reduce greenhouse gas emissions compared to fossil fuels, particularly if the biomass is sustainably sourced and the production process is optimized. However, the availability of sustainable biomass is a key constraint. Large-scale biofuel production can compete with food production, leading to deforestation and other environmental problems. Therefore, it is crucial to use sustainably sourced biomass, such as agricultural residues or algae grown on non-arable land, to avoid these negative impacts.

The transition to sustainable fuel alternatives requires a multi-faceted approach involving technological innovation, infrastructure development, policy support, and collaboration across the maritime industry. Governments can play a crucial role by setting clear emission reduction targets, providing incentives for the adoption of cleaner fuels, and investing in research and development. The development of robust safety standards and regulations is also essential to ensure the safe and reliable operation of vessels using alternative fuels. Ultimately, the widespread adoption of sustainable fuel alternatives will depend on their economic competitiveness with fossil fuels. As technologies mature and production costs decrease, these fuels will become increasingly attractive to shipowners and operators, paving the way for a cleaner and more sustainable future for the maritime industry. The journey will require embracing new technologies and challenging conventional wisdom, but the benefits - a healthier planet and a more sustainable industry - are well worth the effort.

7.4 Advanced Materials in Propulsion

The relentless pursuit of enhanced performance, increased durability, and reduced environmental impact has driven a revolution in the materials employed in marine propulsion systems. No longer can we rely solely on traditional metals; the demands of modern vessels require materials capable of withstanding extreme conditions, minimizing energy losses, and enabling novel propulsion concepts. This chapter delves into the fascinating realm of these advanced materials, exploring their properties, applications, and potential to reshape the future of marine propulsion.

Consider the challenges faced by components immersed in seawater. Corrosion is a constant threat, slowly but surely degrading materials and compromising structural integrity. Add to this the relentless forces of cavitation, the erosive power of high-speed water flow, and the demanding thermal stresses generated by powerful engines. Traditional materials often fall short in the face of such adversity. This is where advanced materials step in, offering tailored solutions to these specific challenges. High-strength alloys, advanced composites, and specialized coatings are just a few examples of how material science is pushing the boundaries of what's possible in marine propulsion. These materials allow for lighter components, reducing overall vessel weight and improving fuel efficiency. They offer enhanced resistance to corrosion and wear, extending the lifespan of critical components and minimizing maintenance requirements. Furthermore, they enable the design of more efficient and innovative propulsion systems.

One of the most promising areas is the application of high-strength alloys. These alloys, typically composed of steel, nickel, or titanium, are engineered to provide exceptional strength-to-weight ratios. This is particularly crucial in components like propellers, where minimizing weight can significantly reduce energy losses due to inertia. Moreover, these alloys often exhibit superior resistance to corrosion and fatigue, ensuring long-term reliability in harsh marine environments. Titanium alloys, while more expensive than steel, offer unmatched corrosion resistance and are therefore ideal for critical components exposed to seawater. Nickel-based superalloys, on the other hand, are renowned for their ability to withstand extremely high temperatures, making them suitable for use in gas turbines and other high-performance engines. The selection of the appropriate alloy depends on the specific application and the trade-off between cost, performance, and durability. Careful consideration of the operating conditions and the desired lifespan of the component is essential in making the right choice. The precise composition and processing of these alloys are also crucial factors in determining their ultimate performance.

Composite materials are another game-changer in marine propulsion. Typically consisting of a reinforcing fiber embedded in a resin matrix, composites offer a unique combination of properties that cannot be achieved with traditional materials. Carbon fiber composites, for example, are incredibly strong and lightweight, making them ideal for applications where weight reduction is paramount. These composites are increasingly used in propeller blades, drive shafts, and even hull structures, contributing to significant improvements in fuel efficiency and vessel performance. Glass fiber composites, while less expensive than carbon fiber, offer good strength and corrosion resistance, making them suitable for a wider range of applications. The design and manufacturing of composite components are complex, requiring careful consideration of fiber orientation, resin type, and manufacturing process. However, the potential benefits of composites in terms of weight reduction, improved performance, and extended lifespan are undeniable. The ability to tailor the properties of composites by varying the fiber type, orientation, and resin matrix allows for the creation of highly optimized components that meet the specific demands of marine propulsion systems.

The role of coatings in protecting materials from corrosion, erosion, and fouling cannot be overstated. Specialized coatings can dramatically extend the lifespan of components exposed to seawater, reducing maintenance costs and improving overall reliability. Anti-corrosion coatings, typically based on epoxy or polyurethane resins, provide a barrier between the metal surface and the corrosive environment. These coatings can be further enhanced with additives that inhibit corrosion or provide sacrificial protection. Anti-fouling coatings prevent the growth of marine organisms on submerged surfaces, reducing drag and improving fuel efficiency. These coatings often contain biocides that inhibit the attachment and growth of barnacles, algae, and other marine life. However, the use of biocides is increasingly regulated due to environmental concerns, leading to the development of more environmentally friendly anti-fouling coatings based on non-toxic materials and surface textures that prevent fouling. Thermal barrier coatings protect components from high temperatures, allowing engines to operate at higher efficiencies. These coatings are typically applied to turbine blades and other components exposed to hot gases, reducing heat transfer and preventing thermal damage. The selection of the appropriate coating depends on the specific application and the environmental conditions.

Looking ahead, the development of new and improved materials will continue to drive innovation in marine propulsion. Nanomaterials, such as carbon nanotubes and graphene, offer the potential for even stronger and lighter composites. These materials are still in the early stages of development, but their potential impact on marine propulsion is enormous. Additive manufacturing, also known as 3D printing, is revolutionizing the way components are designed and manufactured, allowing for the creation of complex geometries and customized materials. This technology enables the production of components with optimized shapes and internal structures, leading to improved performance and reduced weight. Furthermore, additive manufacturing allows for the creation of components with integrated sensors and other functionalities, enabling more intelligent and efficient propulsion systems. The combination of advanced materials and additive manufacturing is opening up new possibilities for the design and manufacture of marine propulsion systems.

Finally, the integration of advanced materials into marine propulsion systems requires a holistic approach that considers the entire lifecycle of the component, from design and manufacturing to operation and maintenance. Careful consideration must be given to the compatibility of different materials, the potential for corrosion and wear, and the ease of repair and replacement. Furthermore, the environmental impact of the materials must be taken into account, with a focus on sustainable materials and manufacturing processes. By embracing a holistic approach, we can unlock the full potential of advanced materials to create more efficient, durable, and environmentally friendly marine propulsion systems. The future of marine propulsion hinges on our ability to innovate and adapt, embracing new materials and technologies that push the boundaries of what's possible. As we navigate the ever-changing seas of technological advancement, the selection and application of advanced materials will undoubtedly remain a critical compass guiding us towards a more sustainable and efficient future.

7.5 AI-Powered Propulsion Control

Imagine a ship slicing through the waves, not just with brute force, but with an almost sentient understanding of the ocean around it. That's the promise of incorporating sophisticated computational methods into propulsion. We are talking about going beyond simple automation and entering the realm of predictive, adaptive, and truly smart propulsion systems. It's about leveraging real-time data and advanced algorithms to make propulsion decisions that optimize fuel consumption, minimize emissions, and enhance overall vessel performance in ways previously thought impossible.

Now, let's consider the traditional approach: a skilled engineer meticulously adjusting engine settings based on experience, intuition, and a handful of sensor readings. While valuable, this approach is inherently limited by human reaction time, cognitive capacity, and the ability to process vast amounts of dynamic data simultaneously. Here's where the computational methods step in. Sophisticated algorithms can analyze data from weather forecasts, wave sensors, GPS, engine performance monitors, and even historical voyage data, all in real-time. This information is then used to make intelligent decisions about optimal speed, propeller pitch, engine load, and even route adjustments. Think of it as having a hyper-efficient co-pilot, tirelessly crunching numbers and optimizing performance every second of the voyage.

One of the most exciting areas lies in predictive maintenance. Traditional maintenance schedules are often based on time intervals or operating hours, regardless of the actual condition of the equipment. This can lead to unnecessary maintenance, wasted resources, and even unexpected breakdowns. By analyzing sensor data from the engine and propulsion system, computational methods can identify subtle patterns and anomalies that might indicate impending failures. This allows for proactive maintenance, scheduling repairs before a major breakdown occurs, minimizing downtime and saving potentially huge costs. Moreover, imagine a system that learns from past voyages. The ship develops a memory of how it performs in various conditions, improving its predictive capabilities over time.

Beyond maintenance, the method also offers opportunities for significant fuel savings. Consider a scenario where a vessel encounters changing weather conditions. Instead of relying on the captain's judgment alone, the system can analyze weather forecasts and wave data to determine the most fuel-efficient route and speed. It might suggest a slight course correction to avoid headwinds or reduce wave resistance, resulting in significant fuel savings over the course of a long voyage. It can also optimize engine load, ensuring that the engine is operating at its most efficient point for the given speed and conditions. This real-time optimization can lead to substantial reductions in fuel consumption and emissions, making shipping more sustainable and environmentally friendly.

This goes beyond individual vessels; imagine a fleet of ships connected to a central computational hub. Each ship contributes its data, creating a vast pool of information that can be used to optimize performance across the entire fleet. The hub can identify optimal routes, share best practices, and even predict future demand, allowing for more efficient scheduling and resource allocation. This interconnectedness creates a powerful network effect, where each ship benefits from the collective knowledge and experience of the entire fleet. This collaborative approach has the potential to transform the shipping industry, making it more efficient, sustainable, and responsive to changing market conditions.

But it is important to acknowledge the challenges. Implementing these systems requires significant investment in sensors, data infrastructure, and skilled personnel. Data security and cybersecurity are also critical considerations, as these systems become increasingly reliant on data connectivity. Furthermore, it is essential to ensure that these systems are user-friendly and reliable, providing clear and actionable insights to the crew. Over-reliance on automation can also be a concern, so it is important to strike a balance between human expertise and computational support. It is crucial that the crew remains in control, with the system acting as a powerful decision-support tool rather than a replacement for human judgment.

Looking ahead, the potential of computational methods in marine propulsion is vast. As sensor technology becomes more advanced and affordable, and as algorithms become more sophisticated, we can expect to see even more dramatic improvements in efficiency, safety, and sustainability. Imagine ships that can automatically adapt to changing sea conditions, optimize their routes in real-time, and even communicate with other vessels to coordinate their movements. This future is not just a pipe dream; it is a tangible possibility, driven by the relentless pursuit of innovation in the maritime industry. The integration of computational methods into propulsion is not just a technological advancement; it is a paradigm shift that will reshape the future of shipping.

The path forward requires a multi-faceted approach. It necessitates collaboration between engineers, data scientists, and maritime professionals to develop and implement these technologies effectively. It also requires investment in education and training to ensure that the next generation of maritime professionals is equipped with the skills and knowledge to operate and maintain these advanced systems. Furthermore, it requires a commitment to open innovation and data sharing to accelerate the development and adoption of these technologies across the industry. The journey towards fully smart propulsion is a marathon, not a sprint, but the potential rewards are well worth the effort. As we continue to explore the possibilities, we can expect to see even more remarkable advancements in the years to come, transforming the way ships navigate the world's oceans. The future of marine propulsion is not just about building better engines and propellers; it's about harnessing the power of information to create a more efficient, sustainable, and intelligent maritime industry.

www.ingramcontent.com/pod-product-compliance
Lightning Source LLC
LaVergne TN
LVHW100325210926
848480LV00003B/1258
* 9 7 9 8 3 4 9 5 8 8 0 4 4 *